IDENTIFICATION OF COMMON INDIAN MEDICINAL PLANTS

V.N. NAIK

SCIENTIFIC PUBLISHERS (INDIA)

Published by:

Scientific Publishers (India)
5 A, New Pali Road, P.O. Box 91
Jodhpur 342 001 (India)

Branch Office
Scientific Publishers (India)
4806/24, Ansari Road, Daryaganj
New Delhi - 110 002 (India)

E-mail: info@scientificpub.com
Website: www.scientificpub.com

ISBN: 81-7233-373-0 (10)
978-81-7233-373-7 (13)

Reprint from:
JOURNAL OF ECONOMIC AND
TAXONOMIC BOTANY
ADDITIONAL SERIES NO. 25
ISSN: 0970-3306

Lasertype set: Rajesh Ojha
Printed in India

CONTENTS

PREFACE

Research in any discipline of plant science demands proper identification of the raw material i.e. species as its pre – requisite. Even in the field of taxonomy, the material has to be identified initially and the improved taxonomic interpretation may follow subsequently. This is particularly so, in case of research in medicinal plants, which are then used in the manufacture of life saving medicines. But, unfortunately it is noticed that many practitioners in the Ayurvedic System of Medicine do not care for the proper identification of plant species being employed to heal the diseased patients. As a result, much of the research in Ayurveda in this country proves to be useless. The foundation of plant identification itself is not very sound in our country and the students of II year B.A.M.S., with Drawyagn as one of the compulsary subjects, do not fair-well in the examinations.

Most of the Ayurvedic medicines available in the market today, do not yield the desired effects. Here again the difficulty lies mainly in proper identification of the raw material used in the manufacture of the medicine. Most of the pharmacists are unaware of the purity of the raw material.

In our country, there are very few cultivators of medicinal plants. Most of the cultivators do not know exactly what they are growing. They do not have the reliable source of supply of pure planting material.

Majority of the students of Botany and Ayurveda do not have experience in plant identification, mainly because they are not trained in this discipline.

Considering all the above problems, it was realized that, the persons in the field of Ayurved, Agriculture, Botany, Medicine and Pharmacy together with those in Forestry and Trade should be provided with an easy tool to identify plant materials based on their external features, in a scientific manner.

The present work is an outcome of long-standing experience in the field of Taxonomy and particularly in the Floristic studies.

The entire work is presented in five uneven chapters. The first chapter deals with the botanical terminology used in the identification and its explanation together with suitable illustrations. The second chapter provides a glossary of all such technical terms and their explanation in brief. The third chapter provides a dichotomous bracket key to the identification of plant families and the fourth chapter provides key to the identification of more than 600 Indian plant species used in Ayurvedic medicines. These include more than 350 species prescribed in the syllabus of II year B.A.M.S. The fifth chapter provides a list of all these medicinal plants with Latin, Sanskrit, Marathi, and Hindi names and their parts used in medicine.

In the keys, pairs of contrasting character states are listed serially (serial numbers listed on the left-hand side) and on the right-hand side the numbers indicate the next step to be followed until one reaches the conclusion. The plant families are according to the recent systems of plant classification. They do not correspond exactly to those given by Charak, Sushrut or Wagbhat.

It is hoped, that, this key will help students of Ayurved, Agriculture, Botany, Forestry, Medicine and Pharmacy in identification of entire plant or plant parts.

Aurangabad **V.N. Naik**

LIST OF ILLUSTRATIONS

Continue ....

Continue...........

COLOUR PHOTO INDEX

Coloured Photographs of Medicinal Plants
for
Confirmation of Identified Plants

001	*Andrographis paniculata*	कालमेघ
002	*Barleria prionitis*	सहचर
003	*Crossandra infundibuliformis*	सैरेयक
004	*Hygrophila schulli*	कोकिलाक्ष
005	*Rhinacanthus nasutus*	यूथिपर्णी
006	*Trianthema portulacastrum*	वसु
007	*Celosia argentea*	षितवार
008	*Anacardium occidentale*	काजू
009	*Semecarpus anacardium*	भल्लातक
010	*Centella asiatica*	मंडुकपर्णी
011	*Foeniculum vulgare*	मिश्रेया
012	*Alstonia scholaris*	सप्तपर्ण
013	*Carissa congesta*	करमर्दक
014	*Holarrhena pubescens*	कुटज
015	*Nerium oleander*	करवीर
016	*Acorus calamus*	वचा
017	*Borassus flabellifer*	ताल
018	*Phoenix sylvestris*	विट खर्जुर
019	*Calotropis gigantea*	अर्क
020	*Gymnema sylvestre*	मधुनाषिनी
021	*Tylophora indica*	पित्तवल्ली
022	*Echinops echinatus*	उत्कंटक
023	*Elephantopus scaber*	गोजिव्हा

Continue ....

024	*Sphaeranthus indicus*	मुंडी
025	*Spilanthes oleracea*	अक्कलकारा
026	*Averrhoa carambola*	कर्मरंग
026a	*Dolichandrone falcata*	मेषषृंगी
027	*Oroxylum indicum*	श्योनक
028	*Bixa orellana*	सिंदूरी
029	*Bombax ceiba*	शाल्मली
030	*Ananas comosus*	अननस
031	*Opuntia elatior*	कंथारी
032	*Bauhinia racemosa*	अष्मांतक
033	*Caesalpinia bonduc*	पूतिकरंज
034	*Cassia fistula*	अरग्वध
034a	*Cassia senna*	स्वर्णपत्री
035	*Saraca asoca*	अशोक
036	*Cadaba fruticosa*	हबब
037	*Capparis decidua*	करीर
038	*Capparis sepiaria*	काकादनी
039	*Capparis zeylanica*	व्याघ्रनखी
040	*Crateva adansonii*	वरूण
041	*Celastrus paniculata*	ज्योतिष्मती
042	*Cleome viscosa*	आदित्यभक्त
043	*Garcinia indica*	साराम्ल
044	*Terminalia arjuna*	अर्जुन
045	*Terminalia chebula*	हरितकी
046	*Argyreia nervosa*	वृद्धदारक
047	*Ipomoea sepiaria*	लक्ष्मणा
048	*Ipomoea mauritiana*	क्षीर विदारी
049	*Ipomoea nil*	शंखिनी
050	*Ipomoea pes-caprae*	मर्याद वेल
051	*Operculina turpethum*	निषोत्तर
052	*Kalanchoe pinnata*	पर्णबीज
053	*Dioscorea bulbifera*	वराही कंद
054	*Cordia dichotoma*	बहुवार
055	*Baliospermum solaniifolium*	दंती
056	*Euphorbia nivulia*	स्नूही
057	*Mallotus philippensis*	कंपिल्लक
058	*Phyllanthus amarus*	भूम्यामलकी

Continue ....

059	*Phyllanthus emblica*	आमलकी
060	*Phyllanthus reticulatus*	कृष्णकांबोज
061	*Securinega leucopyrus*	कांटेपुवण
062	*Abrus precatorius*	गुंज
063	*Clitoria ternatea*	अपराजिता
064	*Erythrina suberosa*	परिभद्र
065	*Mucuna pruriens*	कवच बीज
066	*Psoralea corylifolia*	बाकुची
067	*Pueraria tuberosa*	विदारी
068	*Sesbania grandiflora*	अगस्त्य
069	*Uraria picta*	पृश्नीपर्णी
070	*Curculigo orchioides*	तालमुली
071	*Coleus barbatus*	माईनमूळ
072	*Leonotis nepetifolia*	मातीसूळ
073	*Ocimum tenuiflorum*	तुलसी
074	*Cinnamomum camphora*	कर्पूर
075	*Cinnamomum tamala*	तमालपत्र
076	*Cinnamomum verum*	त्वक
077	*Careya arborea*	कुंभी
078	*Aloe vera*	कुमारी
079	*Asparagus racemosus*	शतावरी
080	*Drimia indica*	वनपलांडु
081	*Gloriosa superba*	कलिहारी
082	*Ammannia baccifera*	अग्निपत्री
083	*Lawsonia inermis*	मेंदिका
084	*Woodfordia fruticosa*	धातकी
085	*Abutilon indicum*	अतिबला
086	*Hibiscus cannabinus*	अंबष्ठा
087	*Hibiscus rosa-sinensis*	जपा
088	*Azadirachta indica*	कडुनिंब
089	*Melia azedarach*	महानिंब
090	*Tinospora cordifolia*	गुडुची
091	*Ficus amplissima*	प्लक्ष
092	*Ficus benghalensis*	वड
093	*Ficus hispida*	काकोदुंबर
094	*Ficus microcarpa*	नांद्रुक
095	*Moringa oleifera*	षिग्रु
096	*Myristica* sp.	जातिफल

Continue ....

097	*Nelumbo nucifera*	पदम
097a	*Boerhavia diffusa*	पुनर्नवा
098	*Nymphaea rubra*	रक्त कमल
099	*Oxalis corniculata*	चांगेरी
100	*Pandanus odoratissimus*	केतकी
101	*Argemone mexicana*	सत्यनाषी
102	*Pedalium murex*	गोक्षुर
103	*Hemidesmus indicus*	कृष्णसारिवा
104	*Piper longum*	पिप्पली
105	*Piper nigrum*	मरीच
106	*Plantago ovata*	इषतगोल
107	*Plumbago zeylanica*	चित्रक
108	*Cymbopogon winterianus*	कुतृण
109	*Desmostachya bipinnata*	दर्भ
110	*Portulaca oleracea*	बृहल्लोणी
111	*Clematis heynei*	मोरवा
112	*Catunaregam spinosa*	मदन
113	*Ixora coccinea*	पनकुळ
114	*Neolamarckia chinensis*	कदंब
115	*Citrus aurantifolia*	निंबुक
116	*Madhuca indica*	मधुक
117	*Bacopa monnieri*	ब्राह्मी
118	*Solanum virginianum*	कंटकारी
119	*Strychnos nux-vomica*	विष तिंदूक
120	*Clerodendrun serratum*	भार्ली
121	*Vitex negundo*	निर्गुडी
122	*Alpinia galanga*	कोष्ठ कुलिंजन
123	*Costus speciosus*	केमुक
124	*Guaiacum officinale*	जीवदारू
125	*Tribulus terrestris*	क्षुद गोक्षुर

Andrographis paniculata (कालमेघ)

Barleria prionitis (सहचर)

Crossandra infundibuliformis (सैरेयक)

Hygrophilla schullii (कोकिलाक्ष)

Rhinacanthus nasutus (यूथिपर्णी)

Trianthema portulacastrum (वसु)

Celosia argentea (शितवार)

Anacardium occidentales (काजू)

9

Semecarpus anacardium (भल्लातक)

10

Centella asiatica (मंडुकपर्णी)

11

Foeniculum vulgare (मिश्रेया)

12

Alstonia scholaris (राप्तपर्ण)

13

Carissa congesta (करमर्दक)

14

Holarrhena pubescens (कुटज)

15

Nerium oleander
(करवीर)

16

Acorus calamus
(वचा)

17

Borassus flabellifer
(ताल)

18

Phoenix sylvestris
(वीट खर्जुर)

19

Calotropis gigaantea
(अर्क)

20

Gymnema sylvestre
(मधुनाशिनी)

Tylophora indica (पित्तवल्ली)

Echinos echinatus (उत्कंटक)

Elephantopus scaber (गोजिव्हा)

Sphaeranthus indicus (मुंडी)

Spilanthes oleracea (अक्कलकारा)

Averrhoa carambola (कर्मरंग)

Dolichandrone falcata (मेषशृंगी)

27 *Oroxylum indicum* (श्योनक)

28 *Bixa orellana* (सिंदूरी)

29 *Bombax ceiba* (शाल्मली)

30 *Ananas comosus* (अननस)

31 *Opuntia elatior* (कंथारी)

32 *Bauhinia racemosa* (अश्मांतक)

33 *Caesalpinia bonduc* (पूतिकरंज)

34 *Cassia fistula* (अरग्वध)

34A *Cassia senna* (स्वर्णपत्री)

Saraca asoca (अशोक)

Cadaba fruticosa (हबब)

Capparis decidua (करीर)

Capparis sepiaria (काकादनी)

Capparis zeylanica (व्याघ्रनखी)

Crateva adansonii (वरूण)

Celastrus paniculata (ज्योतिष्मती)

Cleome viscosa (आदित्यभक्त)

Garcinia indica (साराम्ल)

Terminalia arjuna (अर्जुन)

Terminalia chebula (हरितकी)

Argyreia nervosa (वृद्धदारक)

Ipomoea sepiaria (लक्ष्मणा)

Ipomoea mauritiana (क्षीर विदारी)

Ipomoea nil (शंखिनी)

Operculina turpethum (निशोत्त्र)

Ipomoea pes-capave (मर्याद वेल)

Kalanchoe pinnata (पर्णबीज)

53

Dioscorea bulbifera (वराही कंद)

54

Cordia dichotoma (बहुवार)

55

Baliospermum solanifolium (दंती)

56

Euphorbia niculia (स्नूही)

57

Mallotus philippensis (कंपिल्लक)

58

Phyllanthus amarus (भूम्यामलकी)

59

Phyllanthus emblica (आमलकी)

60

Phyllanthus reticulatus (कृष्णकांबोज)

61

Securinega leucopyrus (कांटेपुवण)

62

Abrus precatorius (गुंज)

64

Erythrina suberosa (परिभद्र)

63

Clitoria ternatea (अपराजिता)

65

Mucuna pruriens (कवच बीज)

66

Psoralea corylifolia (बाकुची)

Pueraria tuberosa (विदारी)

Sesbania grandiflora (अगस्त्य)

Uraria picta (पृश्नीपर्णी)

Curculigo orchioides (तालमुली)

Uoleus barbatus (माईनमूळ)

Leonotis nepetifolia (मातीसूळ)

Ocimum tenuiflorum (तुलसी)

74 *Cinnamomum camphora* (कर्पूर)

75 *Cinnamomum tamala* (तमालपत्र)

76 *Cinnamomum verum* (त्वक)

77 *Careya arborea* (कुंभी)

78 *Aloe vera* (कुमारी)

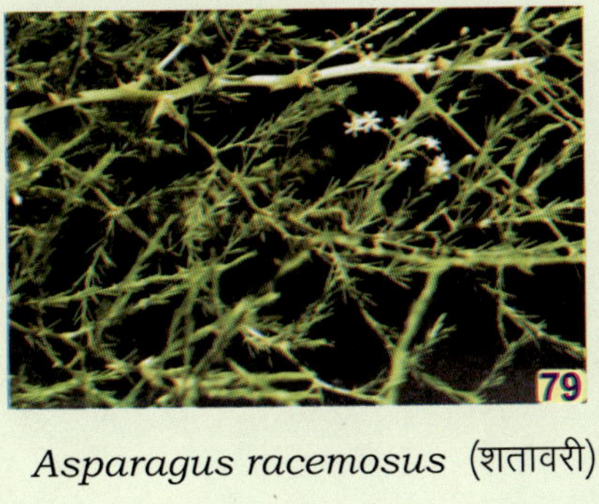

79 *Asparagus racemosus* (शतावरी)

80 *Drimia indica* (वनपलांडु)

81 *Gloriosa superba* (कलिहारी)

82 *Ammannia baccifera* (अग्निपत्री)

84 *Woodfordia fruticosa* (धातकी)

83 *Lawsonia inernis* (मेंदिका)

86 *Hibiscus cannabinus* (अंबष्ठा)

85 *Abutilon indicum* (अतिबला)

87 *Hibiscus rosa-sinensis* (जपा)

Azadirachta indica (कडुनिंब)

Melia azedarach (महानिंब)

Tilospora cordifolia (गुडुची)

Ficus amplissima (प्लक्ष)

Ficus benghalensis (वड)

Ficus hispida (काकोदुंबर)

Ficus microcarpa (नांद्रुक)

Moringa oleifera (शिग्रु)

Myristica sp. (जातिफल)

Nelumbo nucifera (पदम)

Boerhavia diffusa (पुनर्नवा)

Nymphaea rubrs (रक्त कमल)

Oxalis corniculata (चांगेरी)

Pandanus odoratissimus (केतकी)

Argemona mexicana (सत्यनाशी)

Pedalium murex (गोक्षुर)

Hemidesmus indicus (कृष्णसारिवा)

Piper longum (पिप्पली)

Piper nigrum (मरीच)

Plantago ovata (इशतगोल)

Plumbago zeylanica (चित्रक)

Cymbopogon winterianus (कुतृण)

Desnostachya bipinnata (दर्भ)

Portulaca oleracea (बृहल्लोणी)

Clematis heynei (मोरवा)

Catunaregam spinosa (मदन)

Ixora coccinia (पनकुळ)

Neolamarckia chinensis (कदंब)

Citrus aurantifolia (निबुंक)

Madhuca indica (मधुक)

Bacopa monnieri (ब्राह्मी)

Solanum virginianum
(कंटकारी)

Strychnos nux-vomica
(विष तिंदूक)

Clerodendrun serratum
(भार्ली)

Vitex negundo (निर्गुडी)

Alpinia galanga (कोष्ठ कुलिंजन)

Costus speciosus (केमुक)

Guaiacum officinale (जीवदारू)

Tribulus terrestris (क्षुद गोक्षुर)

1

KNOWING PLANTS BY THEIR EXTERNAL FEATURES

INTRODUCTION

Getting acquainted with the common plants growing around is a relatively easy job as their number is relatively small and knowing them does not involve any scientific procedure. This knowledge is often acquired through hear - say or day - to - day discussions about plants, but, if one is interested in knowing plants in a scientific way, he has to make some sincere efforts.

Plants can be known by their external features such as the kind of plant, its habit peculiarities of root, stem, leaves, flowers, fruits and seeds. A long list of external features of plants can be prepared but, usually only a few important features are enough to know most of the plants. Plants are named and grouped on the basis of these features. Plants with similar features are grouped together, those with dis-similar features are placed apart. Closely similar plants are included in smaller groups while less similar ones are placed in larger groups. In other words, the smaller the groups, greater are the similarities shared by its members. Larger groups have lesser number of common features shared by their members. The well - known groups of plants, from smaller to larger, are, species, genus, family, order, class and division. Closely similar plants are grouped under a single species, similar species under a genus, similar genera under a family and so on. Referring a given plant to its respective species, species to its respective family and these to their respective higher groups is a scientific activity known as identification, naming and classification. This activity can be accurate provided the observation of external features is carefully done. Thus, for identification and classification of plants, the study of external features of plants is one of the most important pre-requisite. In the following pages, we shall try to know some of the important external features of plants and their probable functions so that we may be able to identity at least some of the plants used in Ayurvedic Medicine.

KINDS OF PLANTS

There are various kinds of plants in the world. Members of all these kinds are useful in medicine. Some plants are green while the others are non - green. Some have highly organized complex plant body while the others are with simple, one celled plant body. Some bear flowers while the others do not produce them. Thus, on the basis of these and other external features plants of the world are grouped into lower cryptogamic plants and higher phanerogamic plants. Cryptogamic plants are either one celled (unicellular) or many - celled but they do not possess flowers. These are further grouped into lower cryptogams like bacteria, viruses, algae and fungi and the higher cryptogams such as bryophytes and pteridophytes. All these cryptogams reproduce either by vegetative means or by means of minute grains called spores. The sex organs, when present are often microscopic and hidden in the plant body. Algae are green unicellular or multicellular plants living in water or wet conditions. They can synthesize their (autotrophic) food with the help of green colouring matter (chlorophyll) present in their body, water, carbon dioxide and sunlight. Some members of algae are used as food, sources of gelatin and algin etc. fungi (singular, Fungus) are non - green, parasitic or saprophytic plants depending for their food on external sources (heterotrophic). Some of them are used in medicine. Bacteria and viruses are unicellular or chemical organisms which are very minute and difficult to see with the ordinary magnifying glasses. Most of them are disease causing organisms while some are extremely useful to the mankind. Bryophytes are non - flowering plants of wet situations, green, multicellular and reproducing by vegetative as well as sexual means. A few of them are known to possess medicinal properties. Pteridophytes are higher cryptogams with well-organized plant body. They possess roots, stem, leaves and spores for reproduction. Some of them are known to possess medicinal properties.

The most important group known for its medicinal value is the one known as flowering plants (phanerogams). Some flowering plants have flowers that are cone - like made-up of several bracts (scale-like leaves), male organs (stamens) and female organs (ovules). They are known as gymnosperms. A large majority of plants covering this earth are flowering plants bearing typical flowers and fruits enclosing seeds. Since, they are the dominant plants of the world, we shall try to know their features in some details,

THE SEED

The life of typical flowering plants begins from the seed. Seed is a propagule which on sprouting (germination) produces a plant. A typical seed (Fig. 1) has the covering or seed coat. This seed coat consists of outer covering (testa) which is often thick and hard. It protects the embryo within it. The inner covering (tegmen) is usually thin. The seed coat has a small opening through which the seed absorbs water when it is soaked. The embryo of the plant lies well protected inside the seed coat and consists of an axis and seed leaf or leaves (Cotyledons). The end of the axis that emerges out of the seed coat on germination is called radicle, which on further growth develops into a root. The opposite end of the axis, which develops into a shoot, is called plumule. Cotyledons are generally thick as they store food material. Some cotyledons are thin and the food material is stored in the seed in the form of an endosperm. Accordingly, the seeds are termed as non - endospermic or endospermic respectively. The number of cotyledons varies in different plants. In pines it is 7 to 9; in sunflower, pea, bean etc. it is two (dicotyledons) and in jowar, maize, onion etc. it is one (monocotyledons) (Fig. 2).

When seeds are sown in the soil and given enough water they germinate, that means, the radicle of the embryo elongates and emerges out from the pore (micropyle) present in the seed coat. It enters the soil and develops into a root. The plumule emerges out of the seed coat and grows in the air, it develops into a shoot. During this process of germination, the cotyledons may come out of the seed coat and appear above ground (epigeal germination) as in bean, castor etc. (Fig. 3) or remain under the soil (hypogeal germination) as in gram, maize etc. (Fig. 4). *All these features such as presence or absence of endosperm, number of cotyledons and mode of germination help in identification and classification of plants.*

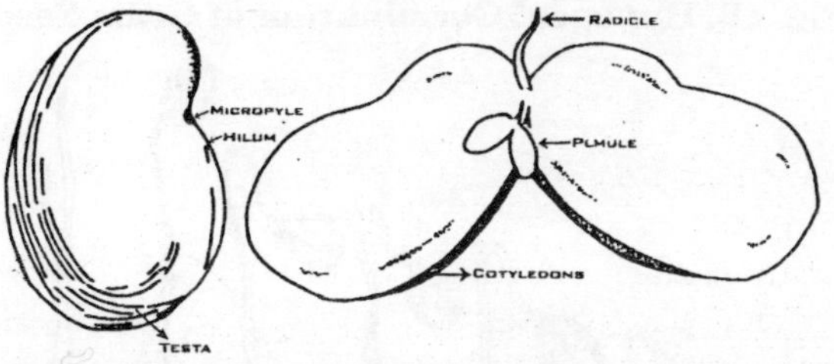

Fig. 1. Bean seed

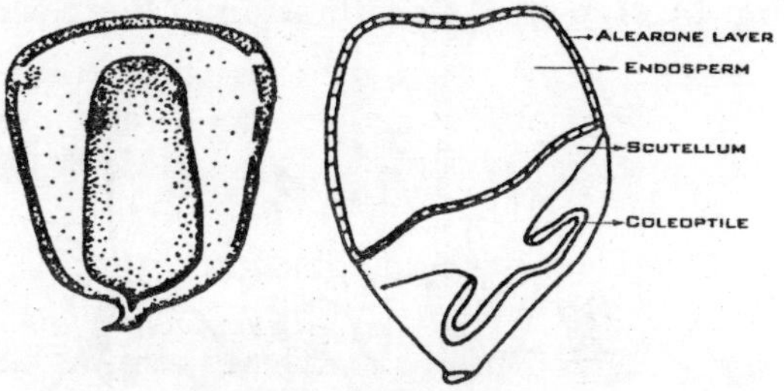

Fig. 2A. Maize grain

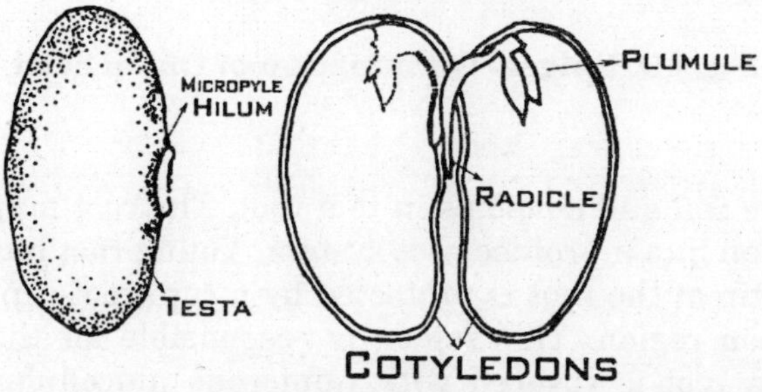

Fig. 2B. Bean Seed

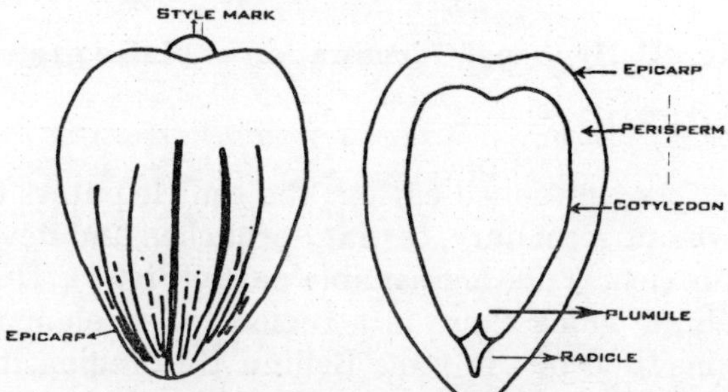

Fig. 2C. Castor Seed

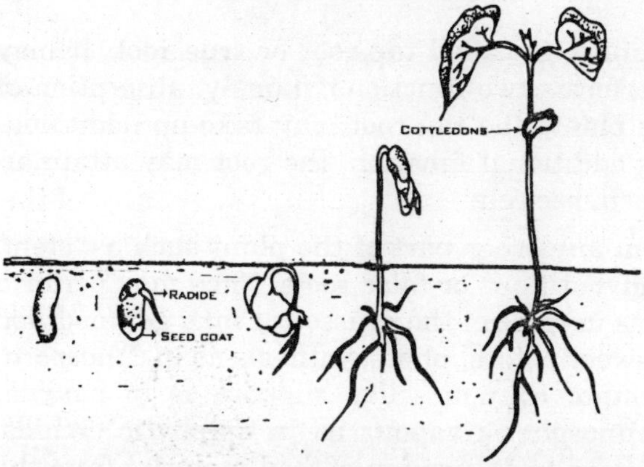

Fig. 3A. Epigeal Germination of Bean Seed

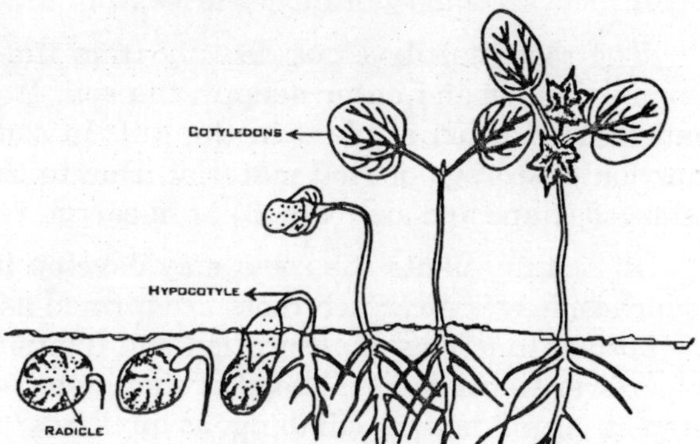

Fig. 3B. Epigeal Germination of Castor Seed

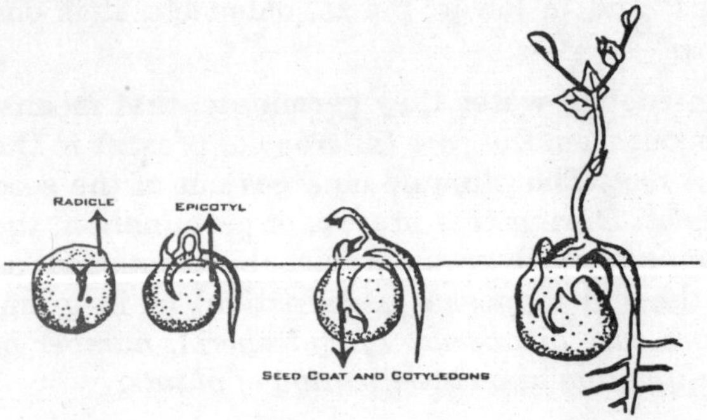

Fig. 4A. Hypogeal Germination of Pea Seed

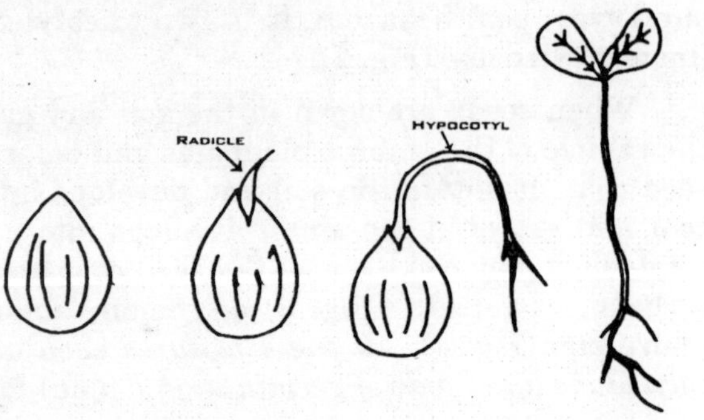

Fig. 4B. Hypogeal Germination of Gram Seed

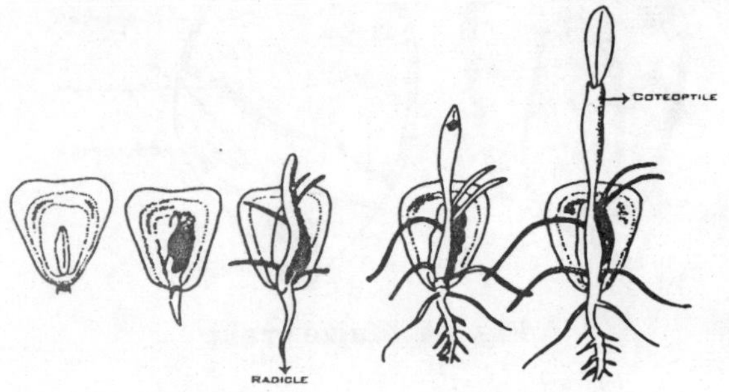

Fig. 4C. Hypogeal Germination of Maize grain

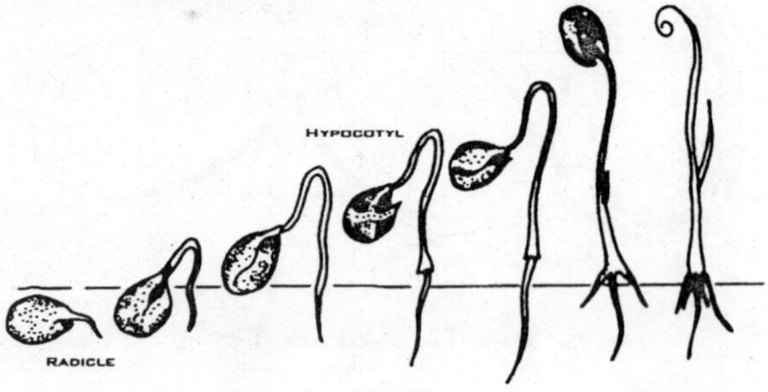

Fig. 4D. Epigeal Germination of Onion Seed

THE ROOTS

As mentioned earlier, the radicle enters the soil and develops in to a root. The root may give out secondary, tertiary branches and develop into a profuse root system. Young root has four clearly distinguishable parts (Fig. 5). The tip of the root is protected by a cap (root-cap) behind which there is a region called elongation region. This region is responsible for the elongation of the root. Behind this region, the root is covered with numerous unicellular hairs (root hairs). This region with root hairs is known as absorption region as the root hairs absorb water from the soil. The region above the absorption region is known as branching region. The secondary branches arise from this region.

The root that develops directly from the radicle is called tap-root or true root. It may remain slender and enter deep in the soil. It performs two functions, namely, absorption of water and fixation of plant in the soil. In some plants the tap-root may take-up additional function of storage of food material. Due to this additional function, the root may attain an enlarged shape and size (Fig. 6) as in carrot, radish, beet etc.

In certain plants the roots may develop from any other part of the plant such as: stem, branches, leaves etc. Such roots are termed as adventitious or false roots. They may perform the normal functions of absorption and fixation as in most of the grasses or may get modified to perform additional functions of storage as in sweet potato, photosynthesis as in *Tinospora*, support as in maize, climbing as in *Ficus pumila,* column - like support as in banyan, respiration as in *Avicennia,* condensation of atmospheric vapour as in epiphytic orchids, fixation of atmospheric nitrogen as in many legumes or absorption of food material from the host plant as in parasites (Fig.7).

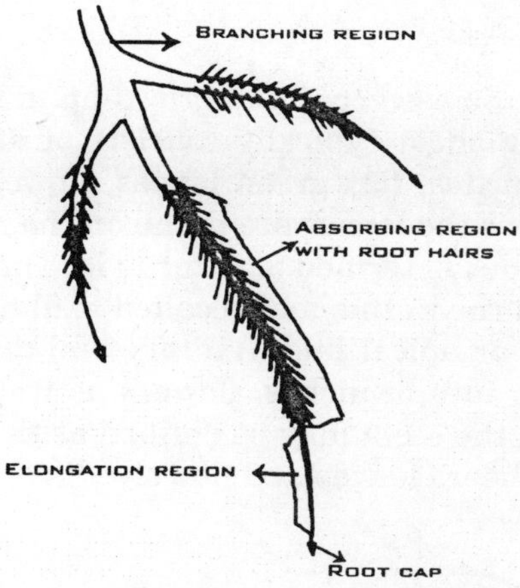

Fig. 5. Parts of a typical Root

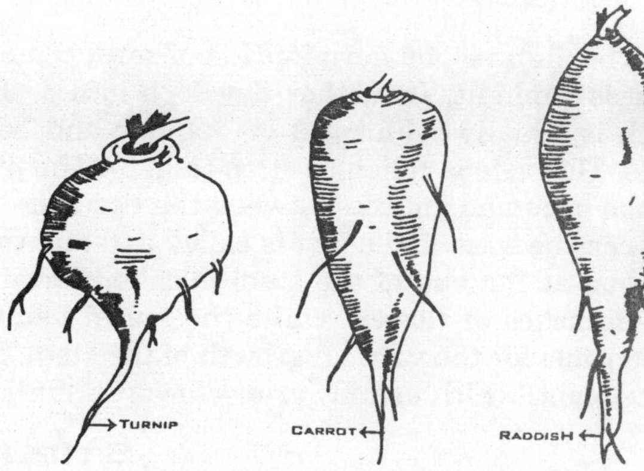

Fig. 6. Forms of Tap Root

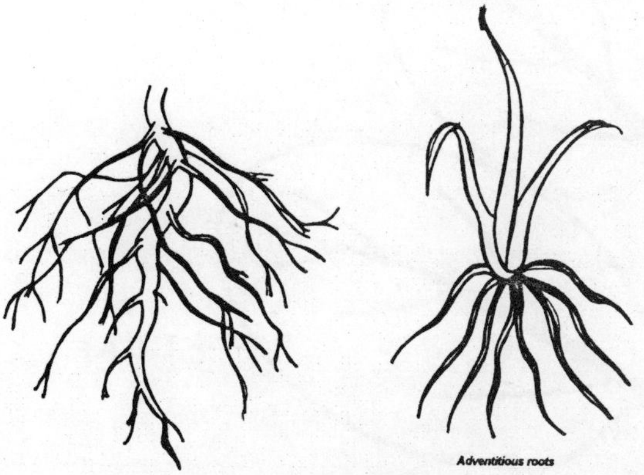

Fig. 7A. Tap Root

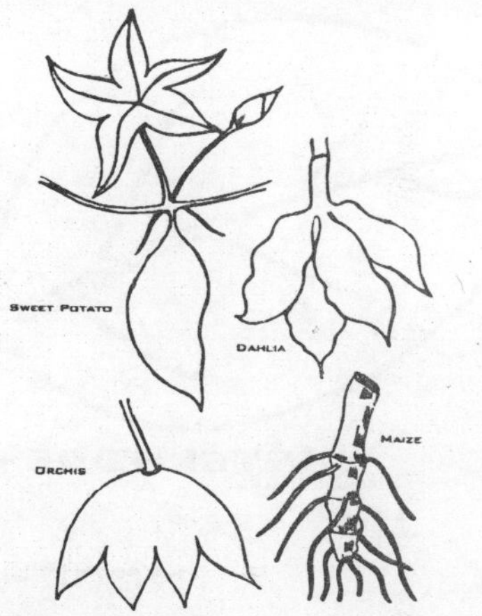

Fig. 7B. Modification of Adventitious Roots

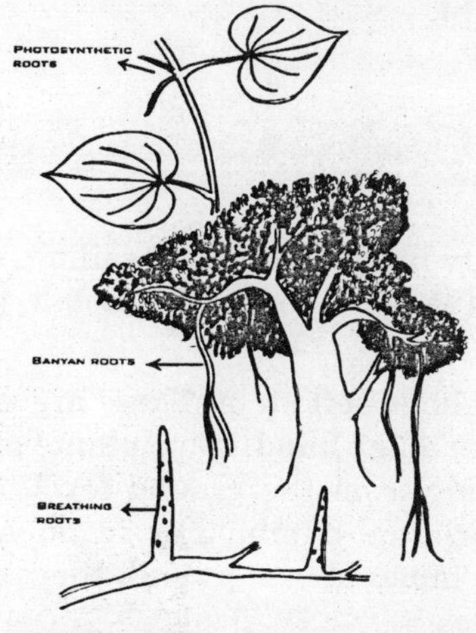

Fig. 7C. Modification of Adventitious Roots

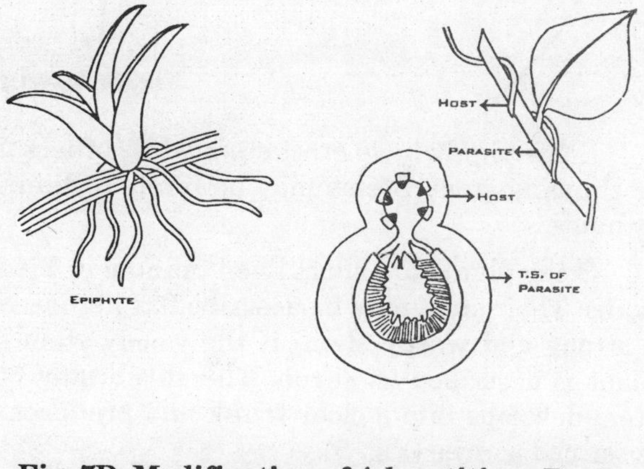

Fig. 7D. Modification of Adventitious Roots

THE STEM

The plumule, on germination of seed, comes out of the seed coat and grows-up in the air towards sunlight. It further develops into a shoot. The shoot typically consists of an axis, which is usually cylindrical or angular and bears branches, flat green leaves, flowers and fruits. This axis is called as stem (Fig. 8). The place where the leaves are borne on the stem is termed node and an axis between the two-successive nodes is termed as internode. The angle between the stem and a leaf is called axil and the bud borne in this axil is called axillary bud. The bud at the end of the stem is called terminal bud or apical bud. Axillary bud develops into branches or flowers while the apical bud develops into branches, flowers as well as is responsible for the vertical growth of the stem. In short, the stem may be defined as that part of the plant, which usually grows towards sunlight and bears leaves and flowers.

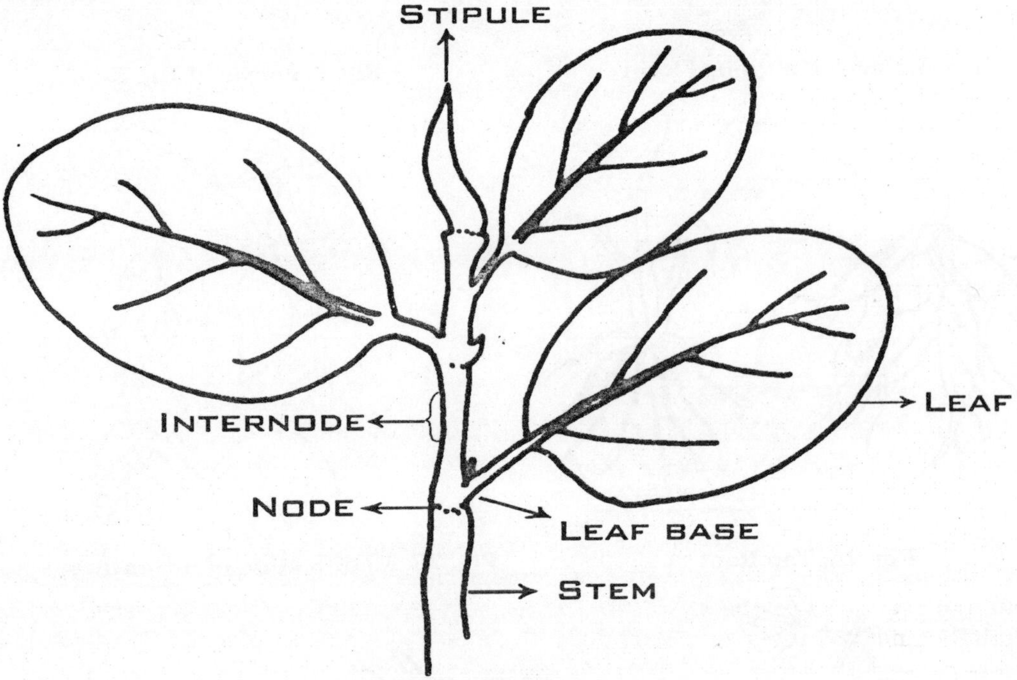

Fig. 8. A twig of Banyan

Some plants have erect stem while others may have twining, scandent, trailing, creeping or climbing stem. Stems may be terete (cylindrical), angular or ribbed and smooth, glabrous or hairy.

Some plants are short lived (annual or biennial) and have soft stem. These are known as herbs. Their stem may be described as herbaceous. On the other hand, some plants may have a strong and woody stem. If the woody stem is branching from the ground level, then the plant is described as shrub. The shrub may be erect, twining, climbing or scandent. If the stem develops into a clear trunk and produces branches much above ground, then the plant is termed as tree (Fig. 9).

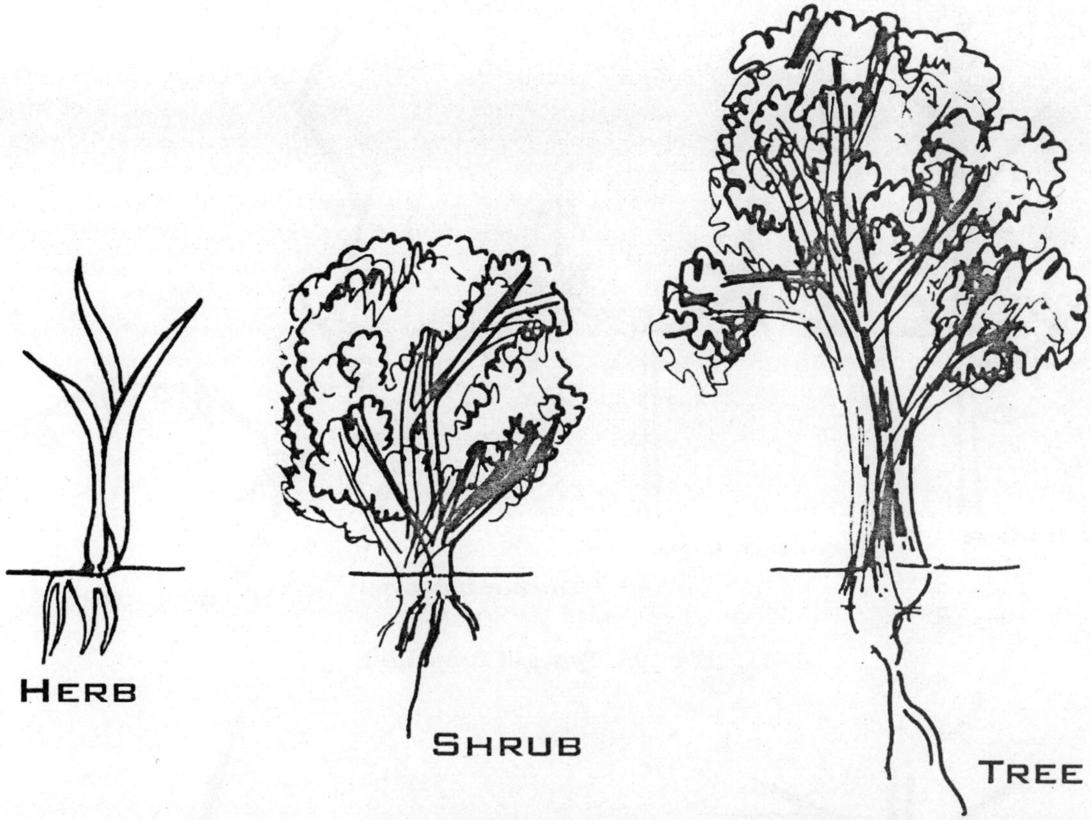

Fig. 9. Habits of Plants

Axillary and terminal buds may develop into branches but different plants have different patterns of branching. May pteridophytes and bryophytes, for instance, show a typical dichotomous branching. Here the terminal bud divides into two buds, each developing into an independent branch. In most of the flowering plants the apical bud continues to grow indefinitely and branches are developed from the axillary buds only. This is called *the racemose pattern of branching*. In certain plants the apical bud develops into a flower and its further growth is stopped. The branches develop from below the flower. This pattern is called as *cymose pattern of branching*. In this pattern, the number of branches produced from below the flower may vary from one (uniparous or monochasial cyme), two (biparous or dichasial cyme) (Fig. 10). In uniparous cymose branching, the successive branches may be produced either on the opposite sides (scorpoid cymose branching) or on the same side (helicoid cymose branching) of the main stem.

The normal functions of stem are —

1. To bear leaves and branches and their proper arrangement;
2. To bear flowers and
3. To distribute water and food material absorbed by roots and leaves respectively to various organs of plants.

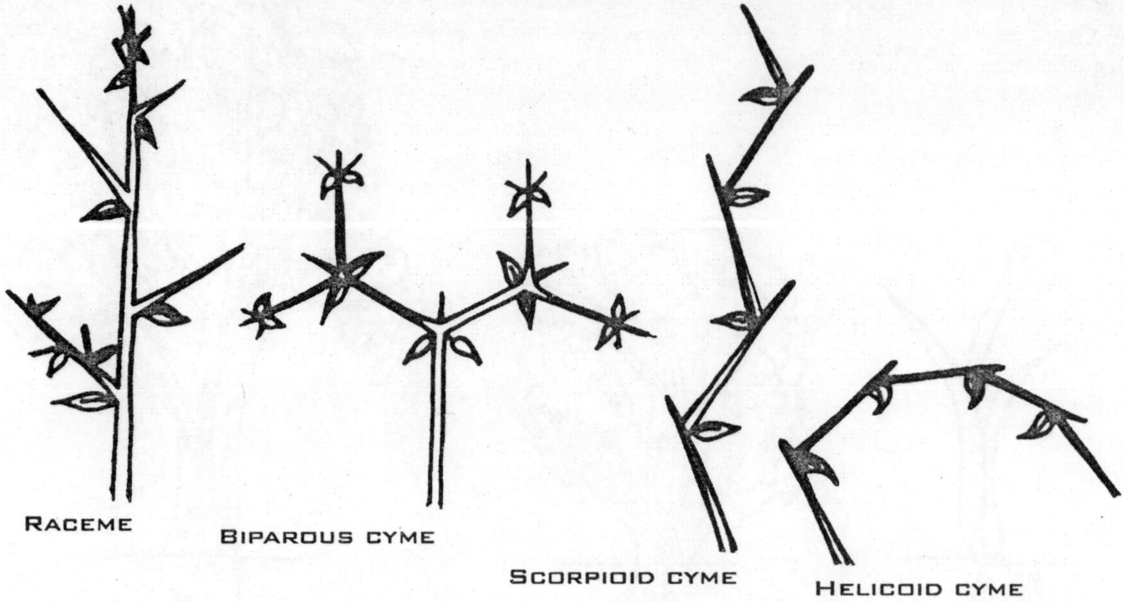

Fig. 10A. Types of Branching

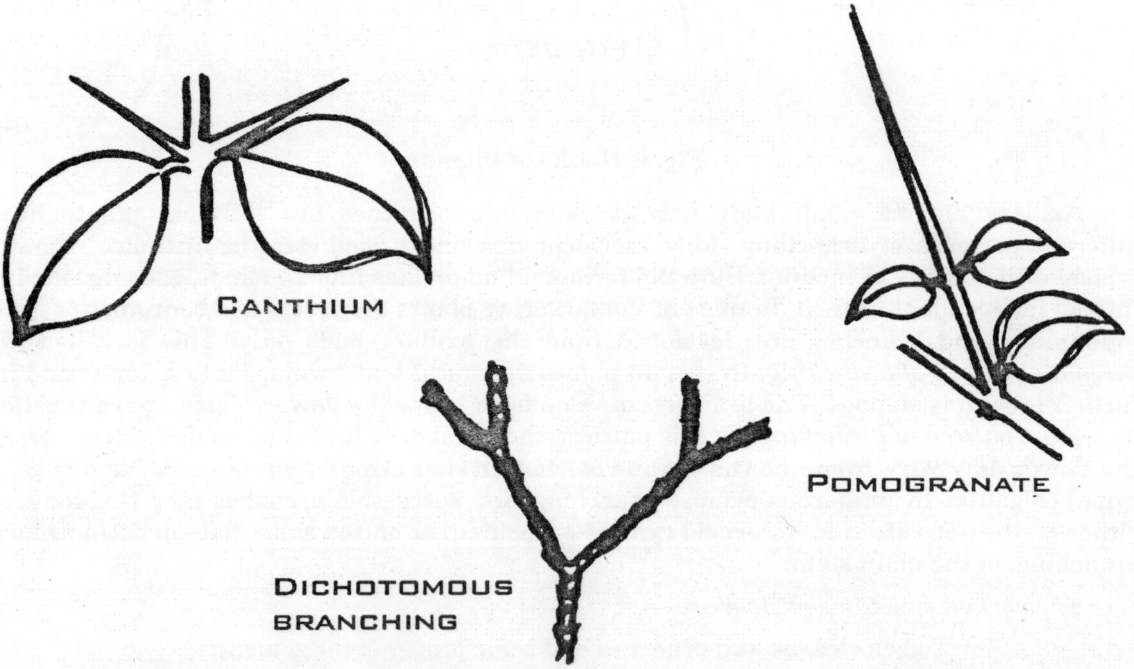

Fig. 10B. Modification of Stem

In addition to these normal functions, the stems of different plants are modified to perform additional functions such as protection (thorns) as in *Carissa,* pomogranate; support (tendrils) as in passion flower, (hooks) as in *Artabotrys*; photosynthesis as in flattened stems of Opuntia (phylloctade) or cladodes (green stems of a single internode) of *Asparagus*; vegetative propagation by means of bulbils as in *Dioscorea, Agave*; spreading or dispersal by means of stolon as in ferns, runner as in *Cynodon*, rhizome as in Banana, offset as in *Pistia* (Fig. 10-18). Certain plants produce under ground stems and store food material in them. These stems perform two fold functions of storage as well as vegetative propagation. They are found in *Curcuma* (rhizome), *Amorphophallus* (corm), potato (tuber) and onion (bulb) (Fig. 19).

Fig. 11. Modification of stem

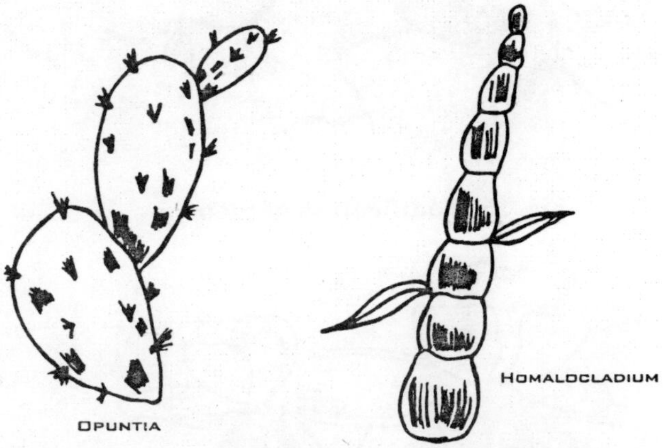

Fig. 12. Modification of Stem-Phylloclades

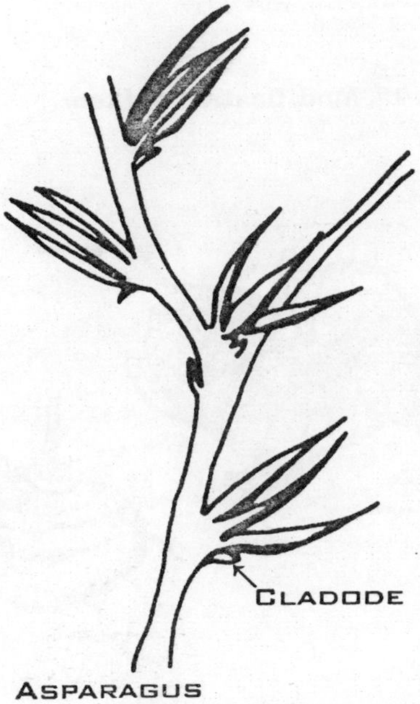

Fig. 13. Modification of Stem

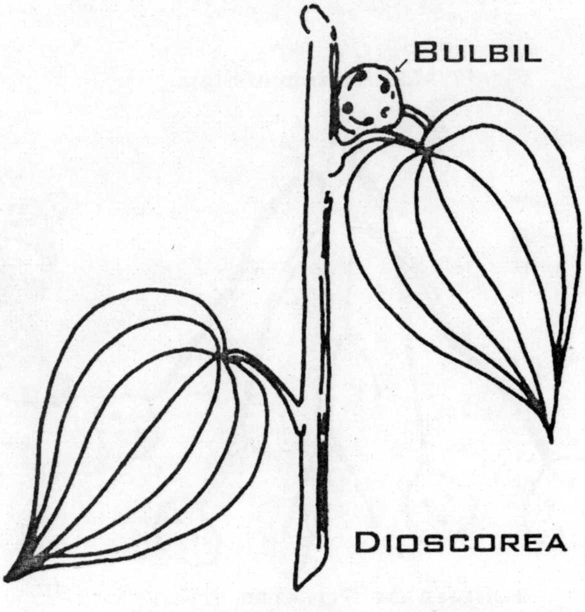

Fig. 14. Modification of Stem

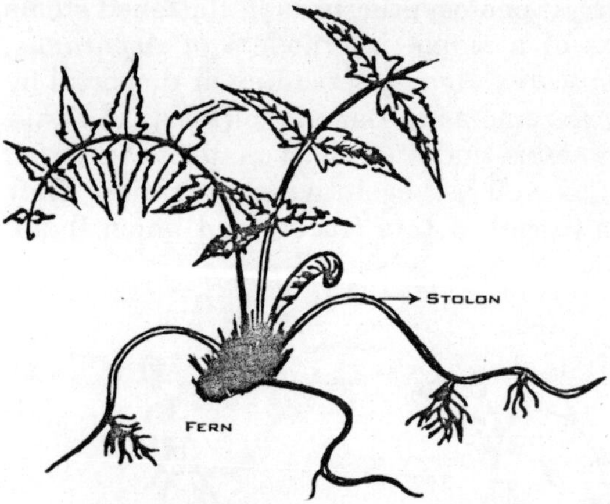

Fig. 15. Modification of Stem

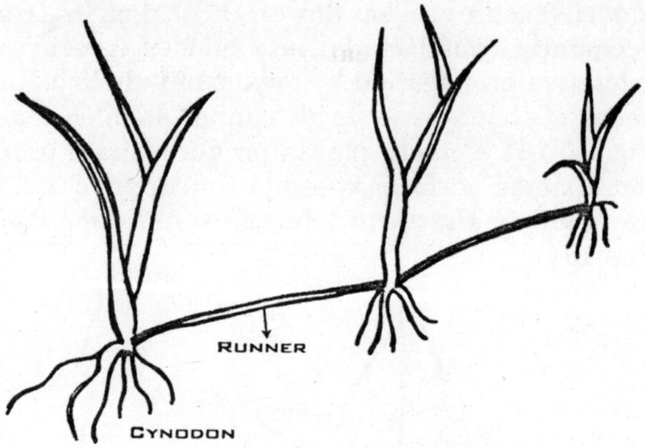

Fig. 16. Modification of Stem

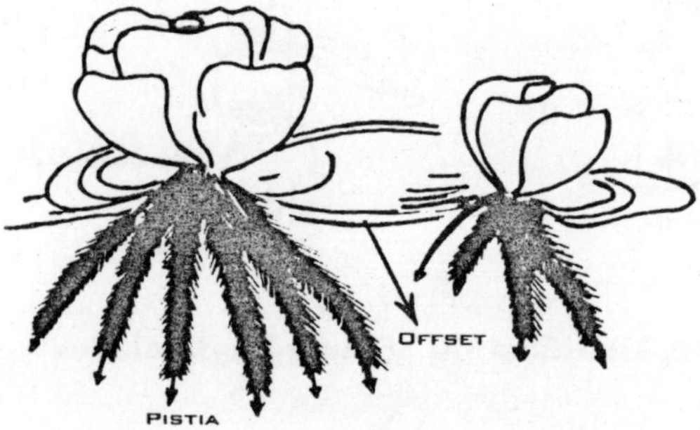

Fig. 17. Modification of Stem

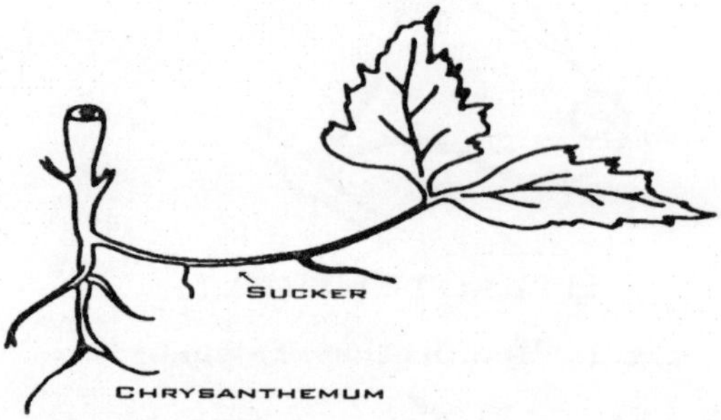

Fig. 18. Modification of Stem

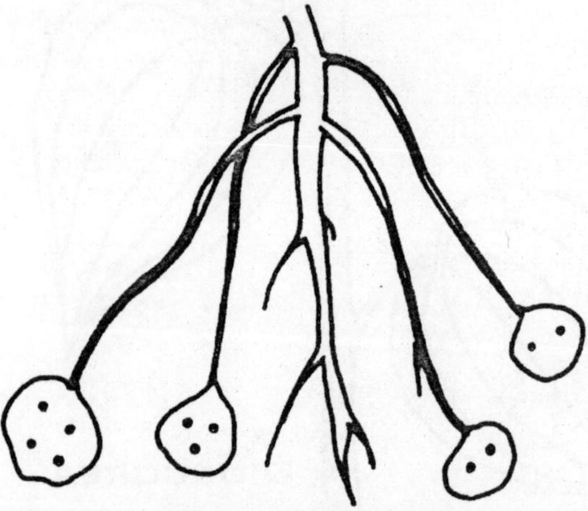

Fig. 19A. Modification of Stem

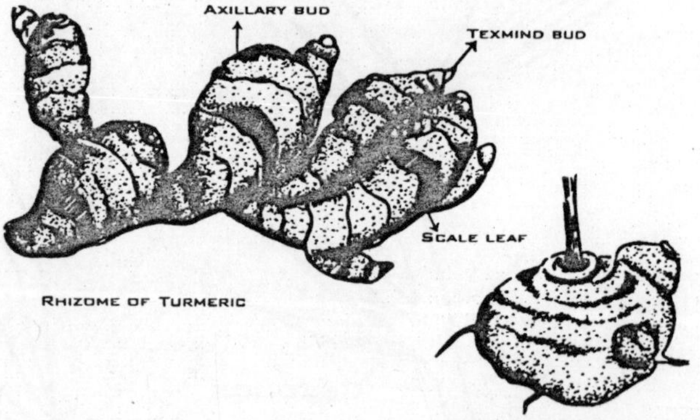

Fig. 19B. Modification of Stem

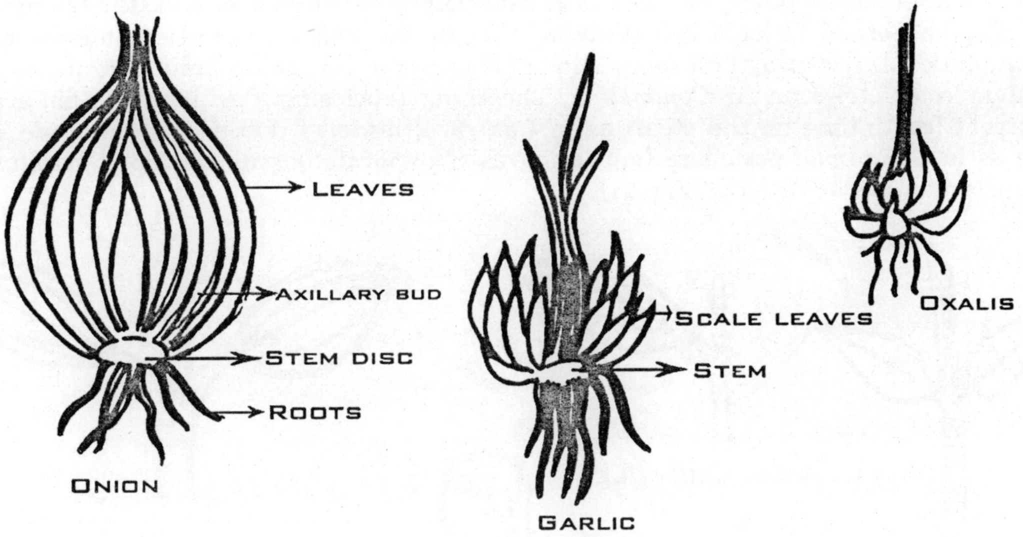

Fig. 19C. Reduced Stems

THE LEAF

Usually the flat, thin and green appendages of the stem are called leaves. The typical leaf has four main parts, namely : 1. Leaf base — the portion where leaf is attached to the stem; 2. Stipules - these are usually two lateral appendages at the base of leaf; 3. Petiole — a stalk attaching leaf to the stem and 4. Leaf blade or lamina — a green, flat, thin portion (Fig. 20). The tip of the lamina is called leaf apex and the sides- leaf margin. In the center of the blade, there is a single strong vein which is called the midrib or main vein and on either side of the midrib, there are numerous side veins and veinlets.

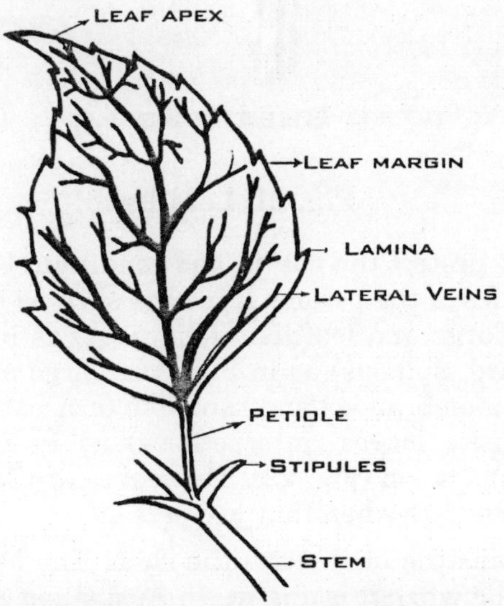

Fig. 20. Parts of a typical leaf

The main function of the leaf base is to bear petiole or if this is absent, the lamina. Leaf base may be normal or enlarged (Pulvinus) as in the leaf of many leguminous plants; semiamplexicaule (covering half the portion of stem) as in coriander; amplexicaule (covering the stem completely) as in *Commelina*; sheathing (enclosing the stem) as in grasses; decurrrent (continuing on the stem) as in *Canscora*; auricled (lobed on either side of the stem) as in *Calotropis*; perfoliate (appearing as if perforated)as in *Cansoora perfoliata* or connate (united) as in *Flaveria* (Fig. 21).

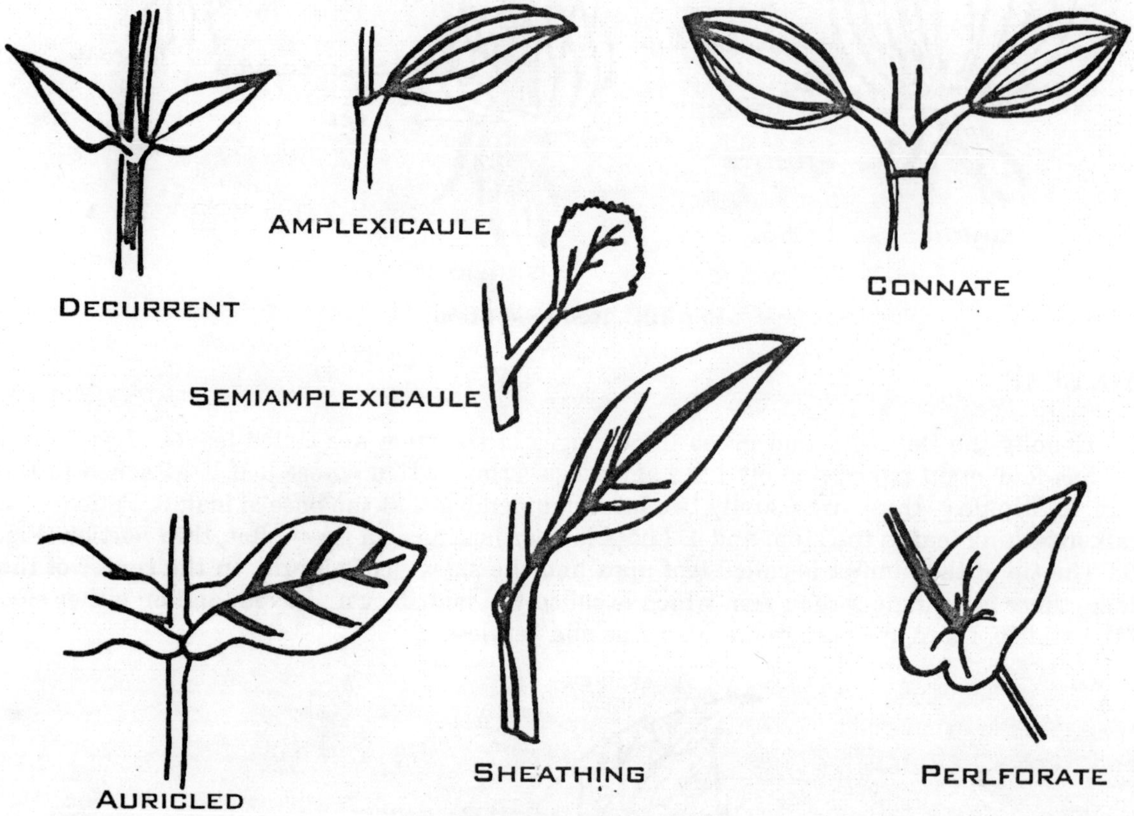

Fig. 21. Leaf Base

The stipules normally protect the leaf in bud condition. Usually they are green, slender and arise on the lateral sides of the petiole. In plants such as rose, they may be connate with the petiole. They may be large and leaf like (foliaceous) as in pea, dry and scale like as in *Alysicarpus*, sharp and hard (spinous) as in *Ziziphus*, large and protecting apical bud (bud-scale) as in banyan, thin, membranous and tube like (ochreate or ocreate) as in *Polygonum*, in between the two opposite leaves (interpetiolar) as in *Ixora*, thread like and coiling (tendrillar) as in *Smilax* and so on (Fig. 22). The leaves are termed as exstipulate when the stipules are absent, and stipulate when they are present.

The petiole usually joins the blade with the stem. The leaf is called petiolate when the petiole is present and sessile when it is absent. In Australian *Acacia*, the petiole becomes flat and appears like a lamina. Such flattened, green petiole is termed as phyllode (Fig. 23).

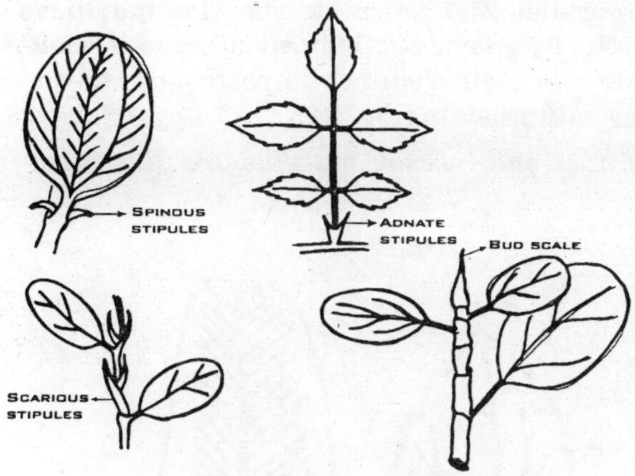

Fig. 22A. Types of Stipules

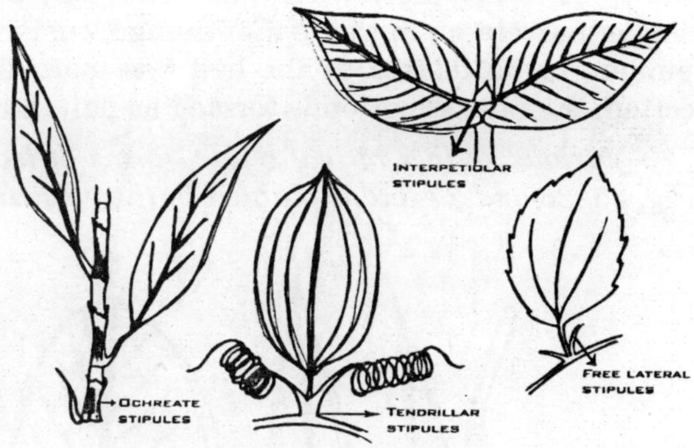

Fig. 22B. Types of Stipules

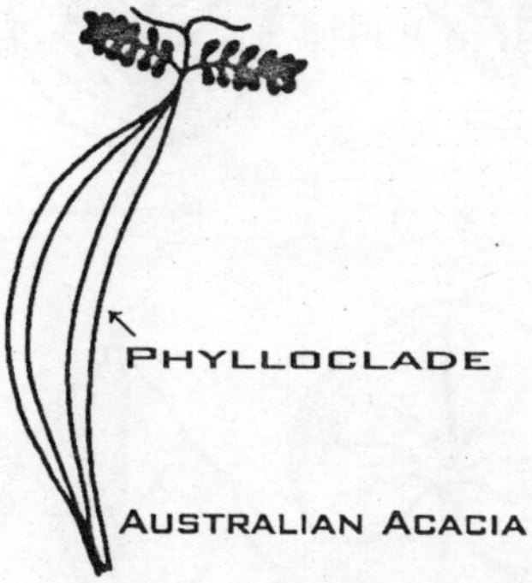

Fig. 23. Modification of Petiole

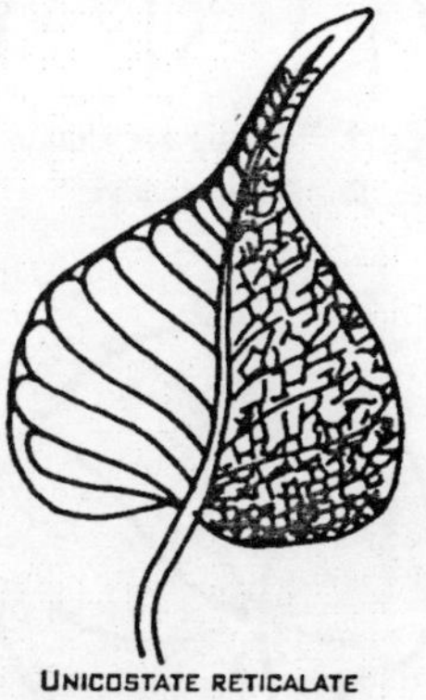

Fig. 24A. Type of Venation

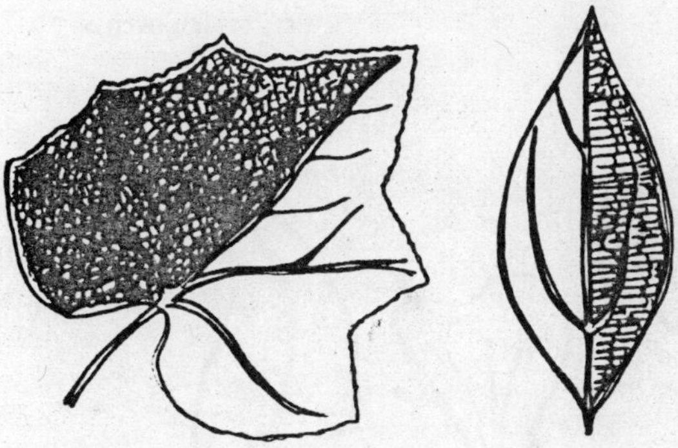

Fig. 24B. Types of Venation

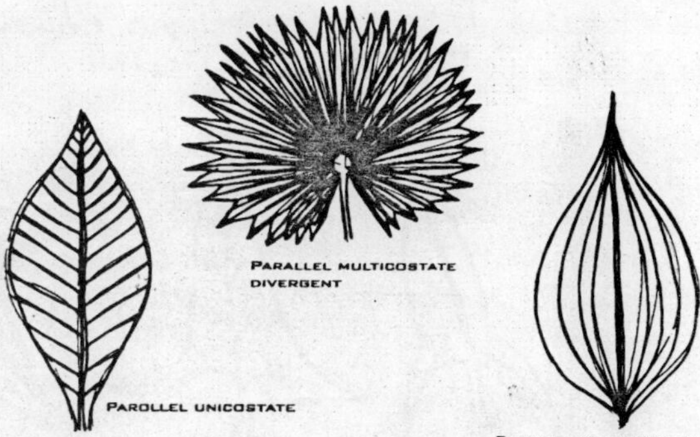

Fig. 24C. Types of Venation

The lamina or blade is the main part of the leaf and is usually thin, flat and green. The veins provide shape to the lamina and are meant for supply of water and transportation of food. The arrangement of veins is called venation. The venation is termed reticulate when the veins form a network or the veins may be parallel. The reticulate venation may have a single midvein as in the leaf of mango, custard apple, guava etc. This venation is termed as pinnate or unicostate. If the leaf has more than one main veins as in castor, cucumber or cotton, then the venation is termed as palmate or multicostate (Fig. 24).

Various shapes of leaf blade, forms of leaf apex and types of margins are illustrated in Fig. 25, 26 and 27 and explained in the glossary.

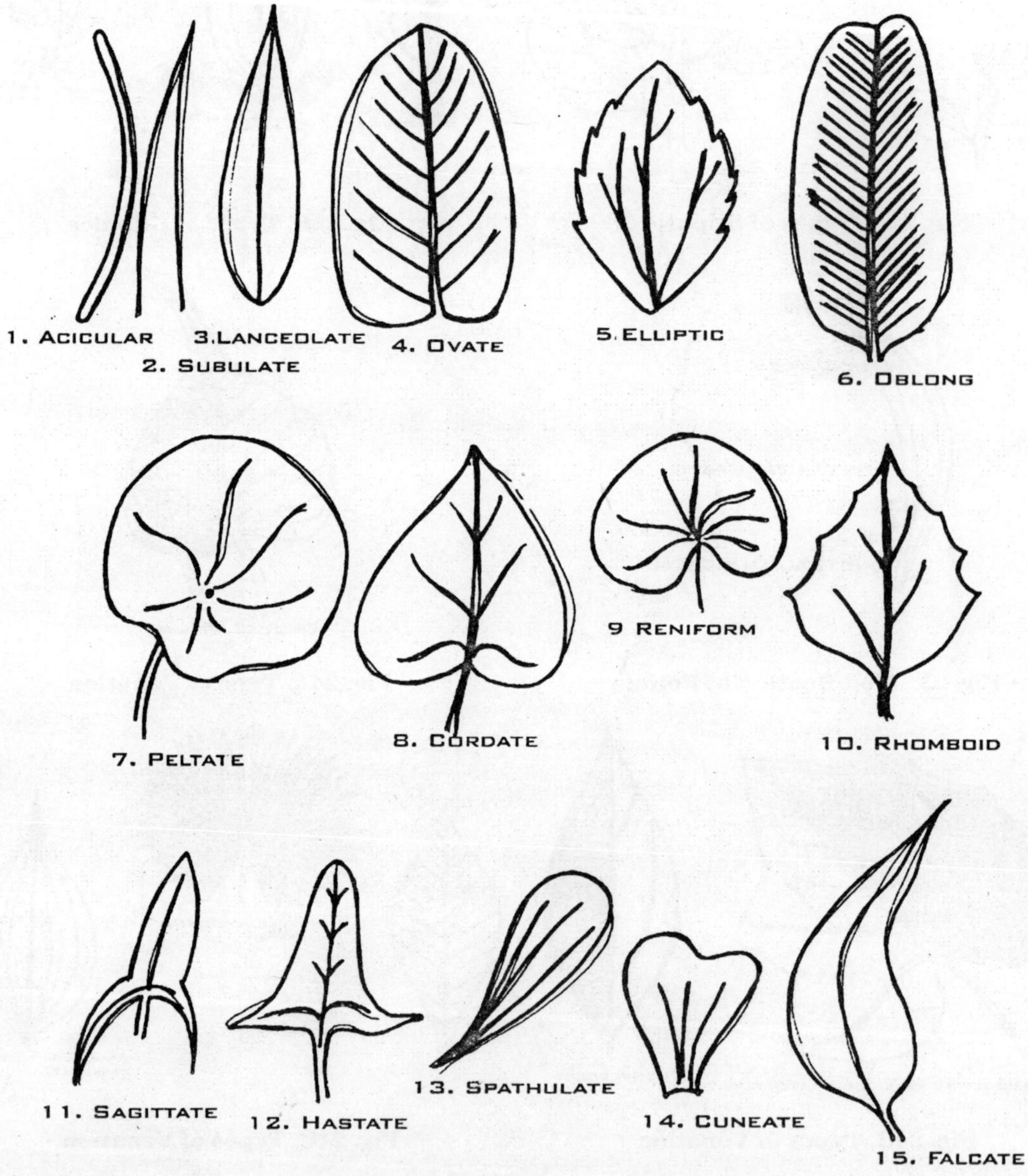

Fig. 25. Leaf shapes

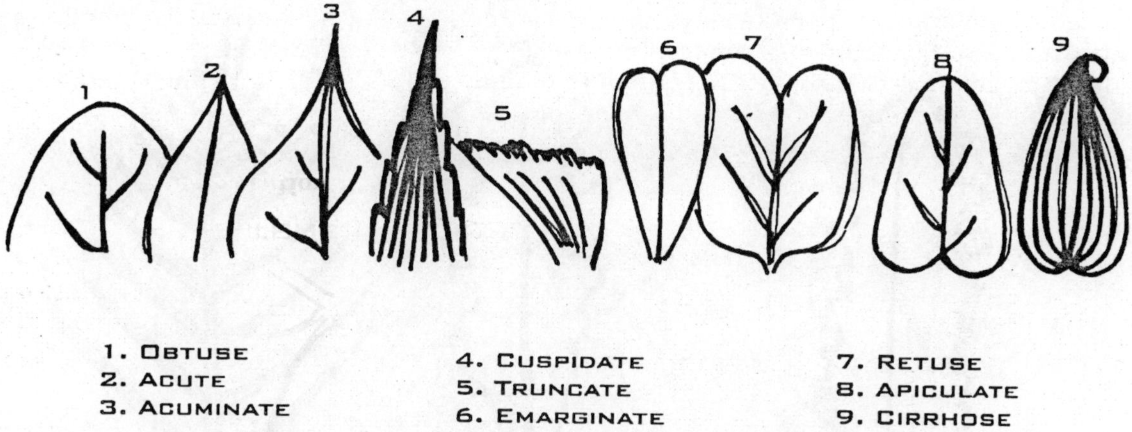

Fig. 26. Leaf Apex

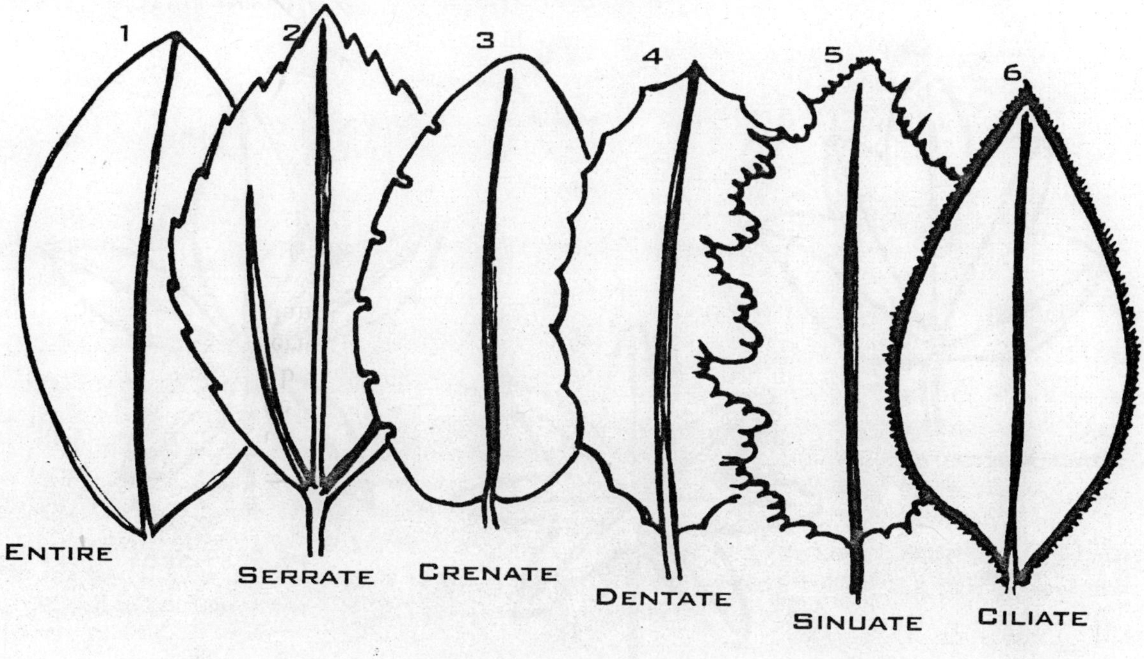

Fig. 27. Leaf Margins

The leaf margin may be entire as in mango, banyan etc. or it may be variously lobed or incised. The incision of the leaves is of two main types namely, that of the pinnate and palmate leaf. In pinnate leaves the lobing may be shallow (pinnatifid) as in *Argemone*, slightly deeper (pinnatipartite) as in radish or much deeper up-to the midrib (pinnatisect) as in coconut. Similarly, the palmate leaf may be shallowly lobed (palmatifid) as in cotton, deeply lobed (palmatipartite) as in castor or much deeply lobed (palmatisect) as in *Hibiscus cannabinus* (Fig. 28).

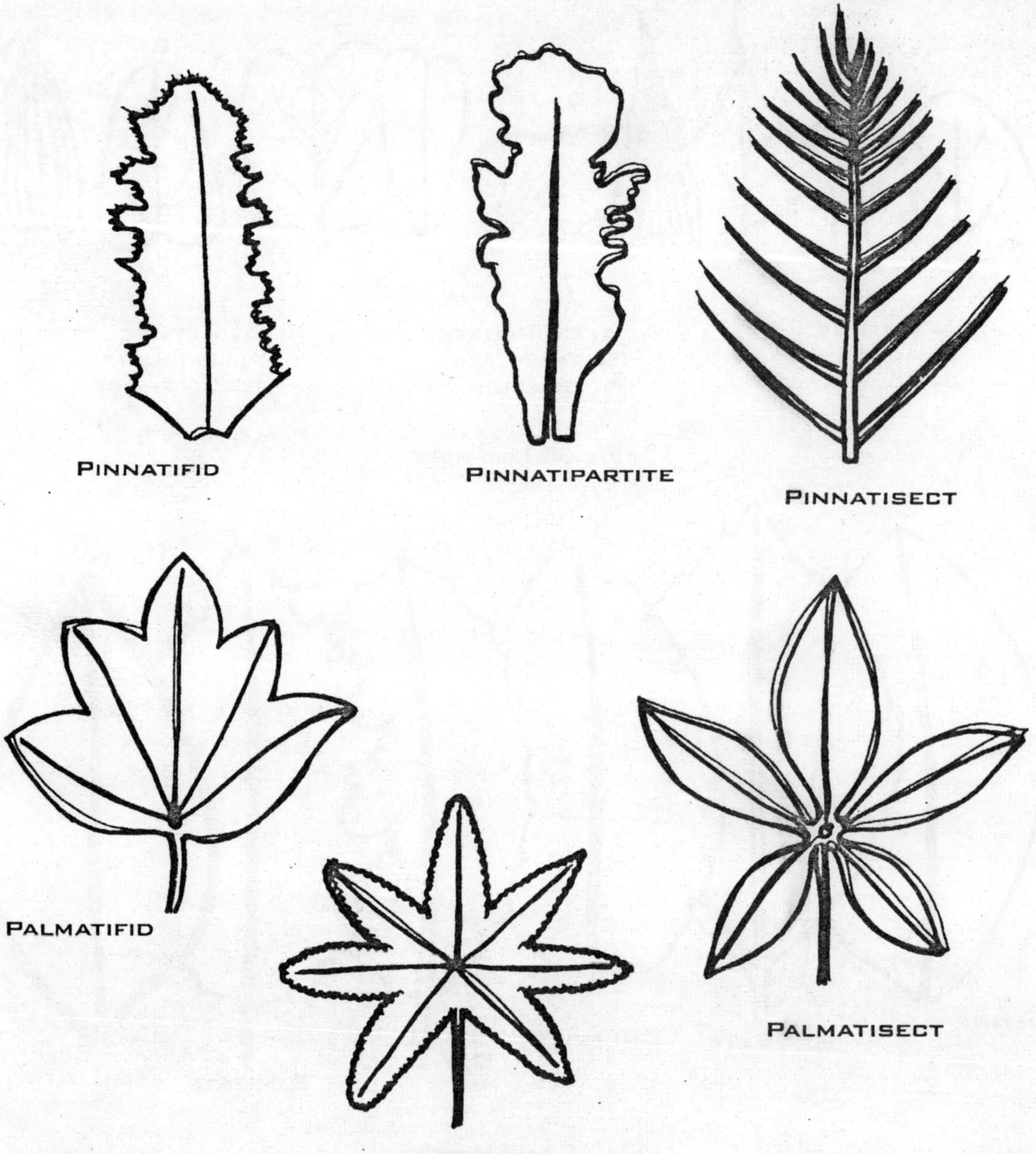

Fig. 28. Leaf Incisions

The leaf is said to be compound when the lobes described above beome completely separate from each other. Each distinct part of such leaves is termed as a leaflet. Each leaflet may either have its own stalk or petiolule or it may be joined to the common petiole. This is the main difference between the simple leaf and compound leaf. Compound leaves are also of two main types, namely, pinnate and palmate compound or digitate leaves.

The leaves of Rose, *Butea, Clitoria* have 3, 5, 7 and other odd numbers of leaflets. They are therefore called imparipinnate leaves. If on the other hand, the number of leaflets is even, then it is called paripinnate leaf as in *Cassia*, Tamarind etc. (Fig. 29).

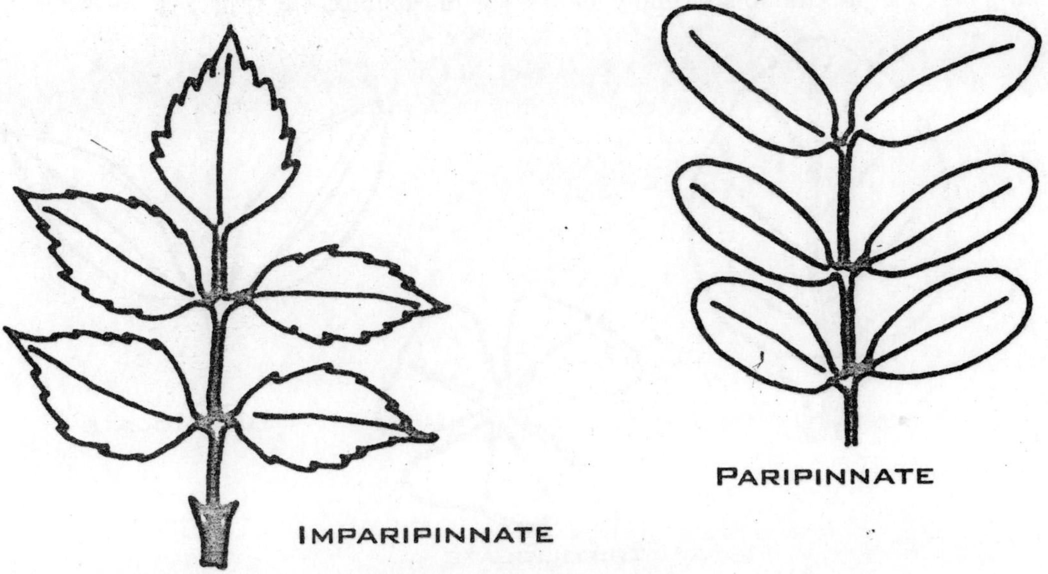

Fig. 29. Compound Leaves

If the pinnate leaf has a simple axis then it is termed as simple pinnate or unipinnate. If the axis is once branched, it is bipinnate, twice branched tripinnate and so on. Rose, *Cassia* etc. are examples of unipinnate, *Acacia* - bipinnate, *Moringa* - tripinnate and *Oroxylum* - multipinnate (Fig. 30).

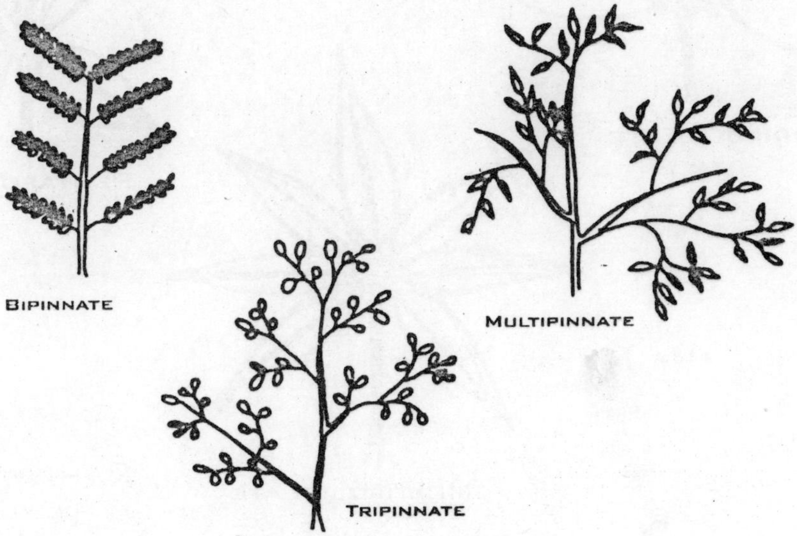

Fig. 30. Compound Leaves

According to the number of leaflets the digitate leaves are termed as unifoliolate (one leaflet) as in lemon, bifoliolate (two leaflets) as in *Balanites*, trifoliolate (three leaflets) as in *Oxalis*), quadrifoliolate (four leaflets) as in *Pithecellobium*, pentafoliolate (five leaflets) as in *Cleome gynandra*, multifoliolate (many leaflets) as in *Bombax*, etc. (Fig. 31).

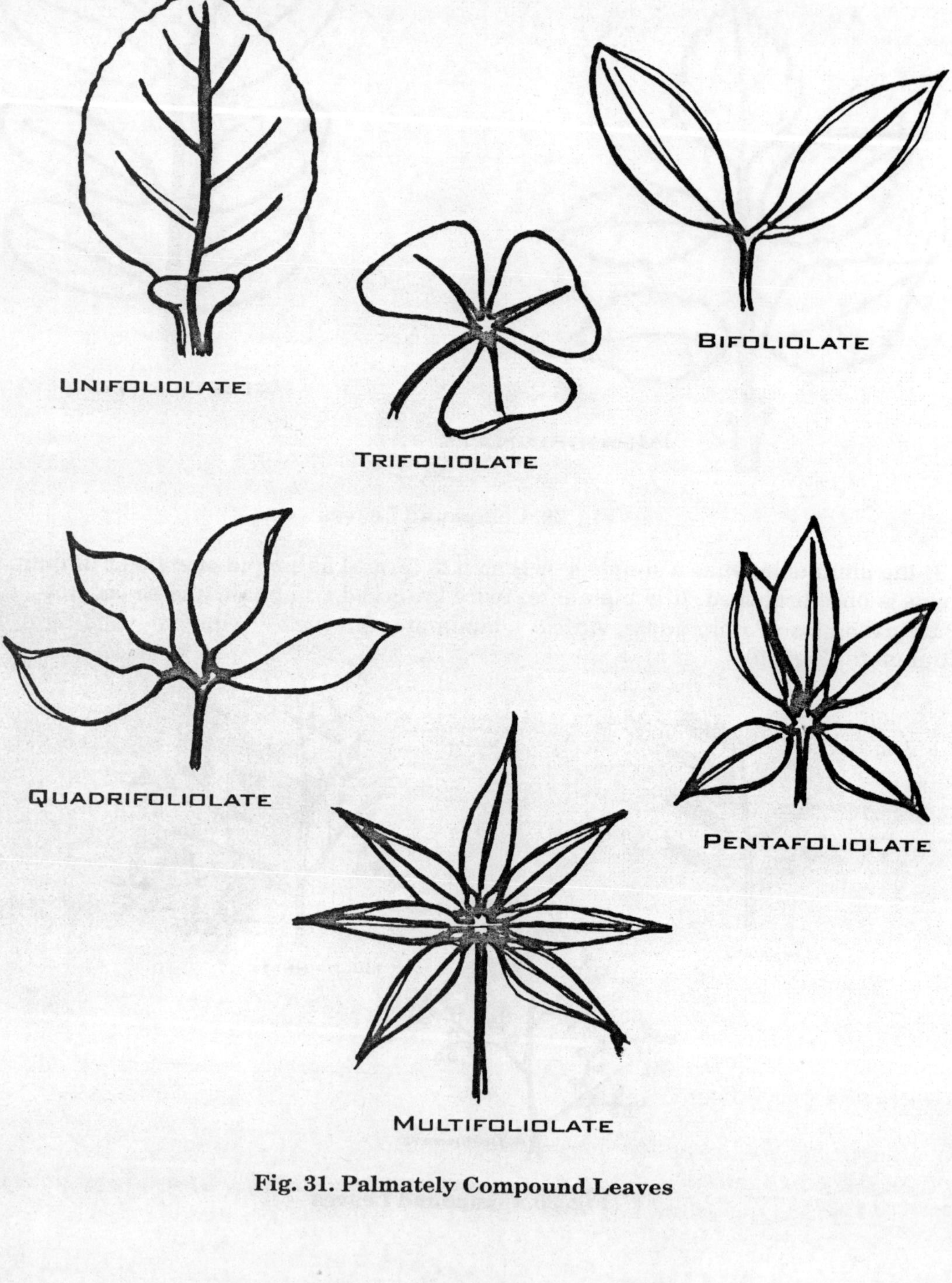

Fig. 31. Palmately Compound Leaves

The arrangement of leaves on the stem is termed as phyllotaxy. There are three main types of phyllotaxy, namely : 1. Alternate or spiral (one leaf at each node) as in Annona, Mango etc. 2. Opposite (two leaves at each node) as in *Calotropis* and 3. Whorled (three or more leaves at each node) as in *Nerium.* The alternate leaves may be arranged in a spiral or in two ranks (distichous) on the stem. Similarly, opposite leaves may be arranged in two ranks (superposed) as in *Quisqualis* or in four ranks (decussate) as in *Calotropis, Ocimum* etc. (Fig. 32).

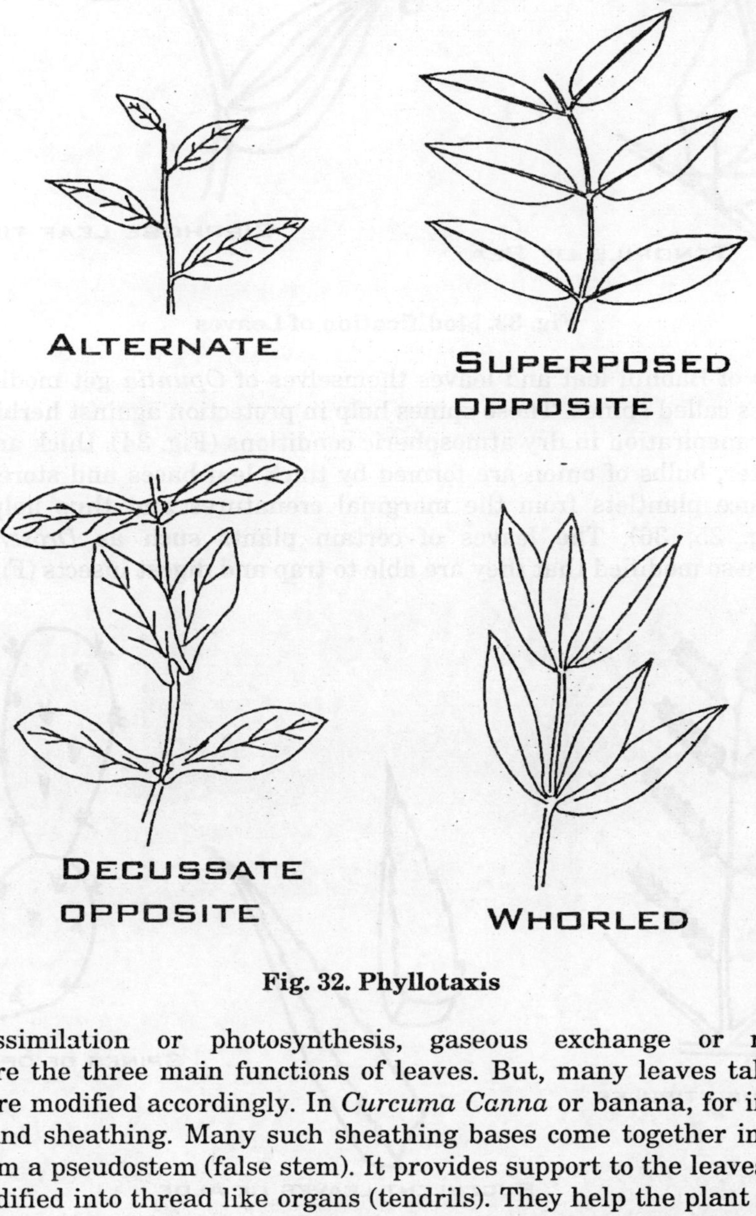

Fig. 32. Phyllotaxis

Carbon assimilation or photosynthesis, gaseous exchange or respiration and transpiration are the three main functions of leaves. But, many leaves take-up additional functions and are modified accordingly. In *Curcuma Canna* or banana, for instance, the leaf bases are flat and sheathing. Many such sheathing bases come together in an overlapping manner and form a pseudostem (false stem). It provides support to the leaves. In *Smifax*, the stipules are modified into thread like organs (tendrils). They help the plant in climbing on a support. In *Gloriosa*, leaf tip is coiled like a thread (cirrhose leaf tip) and helps the plant to climb on the support (Fig. 33).

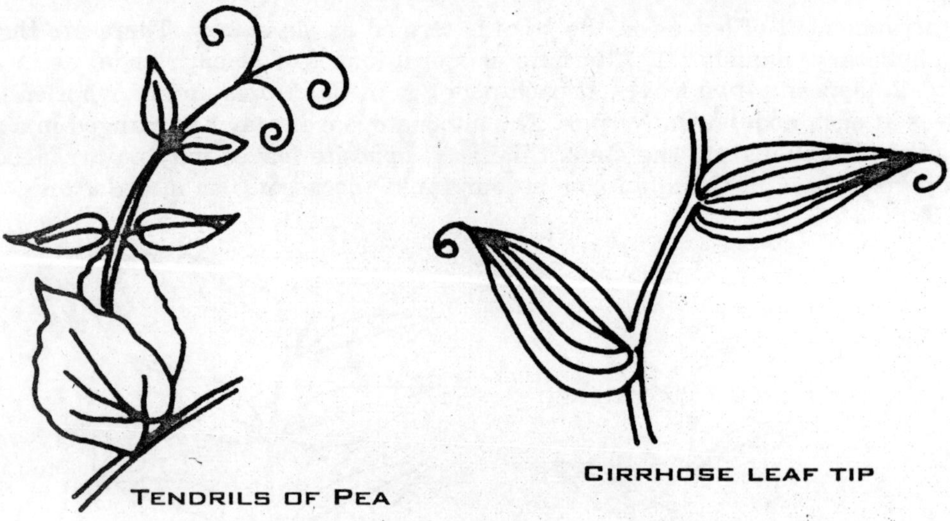

Fig. 33. Modification of Leaves

The stipules of Babhul leaf and leaves themselves of *Opuntia* get modified into sharp, pointed structures called spines. These spines help in protection against herbivorous animals and also check transpiration in dry atmospheric conditions (Fig. 34). thick and fleshy leaves of *Aloe* store water, bulbs of onion are formed by thick leaf bases and store food, leaves of *Kalanchoe*, produce plantlets from the marginal crenatures and thus help in vegetative propagation (Fig. 35, 36). The leaves of certain plants such as *Drosera, Utricularia, Nepenthes* etc. are so modified that they are able to trap and digest insects (Fig. 37).

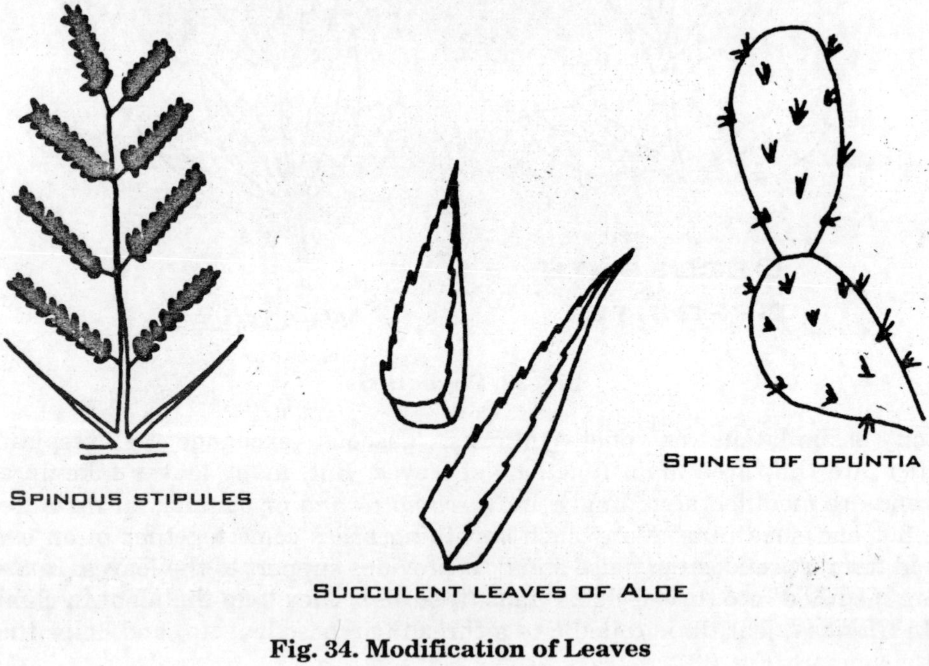

Fig. 34. Modification of Leaves

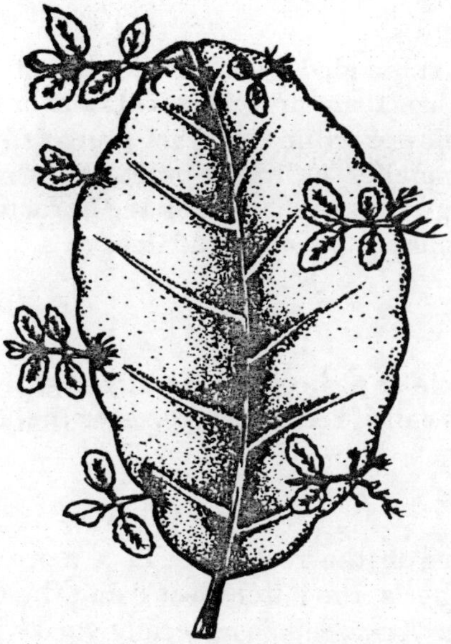

Fig. 35. Leaf of Kalanchoe Bearing Plantlets

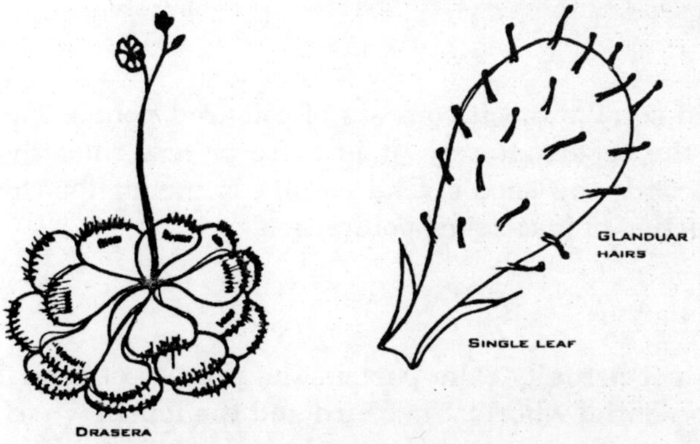

Fig. 36. Leaves of Insectivorous Plants

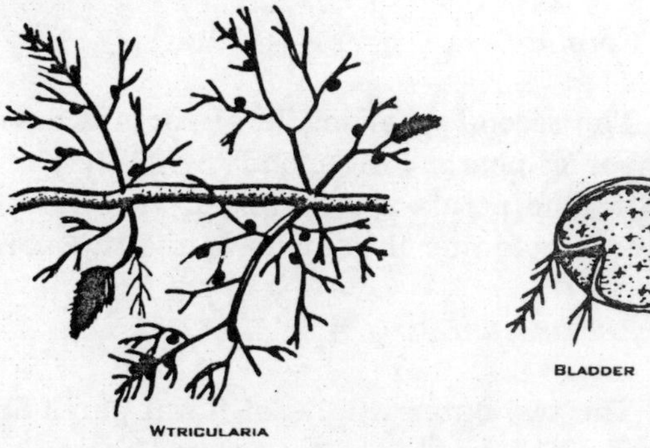

Fig. 37. Insectivorous Plant

THE FLOWER

The root, stem and leaves are the vegetative organs of a plant. The flower is the reproductive organ. The sexual reproduction in plants takes place through the flower.

The typical flower has the following parts:

The peduncle

This is the stalk of the individual flower. If the flowers are arranged in a bunch or in inflorescence, then the common stalk of all the flowers is called the *peduncle,* where as, the stalk of the individual flower is called the *pedicel*. The flowers with pedicels are called pedicellate flowers and when they are without pedicels, they are termed sessile.

The Bract

The flowers are generally subtended by a leaf-like or minute structure called bract. The flowers with bracts are called bracteate and those without bracts are called ebracteate. The peduncle or pedicel bears the flower while the bract protects it in bud condition. In addition to the bract, there are two smaller or inner bract-like structures, these are called the bracteoles and flowers bearing them are bracteolate. Ebracteolate flowers lack bracteoles. The bracteoles provide additional protection to the flower.

The receptacle or torus

The free end of the peduncle or pedicel is generally slightly or considerably enlarged and bears other floral parts. This is called receptacle, torus or thalamas.

The Calyx

The floral parts are borne on the receptacle in a more or less whorled manner. The typical flower has four such whorls, the outer most being the calyx. Calyx is generally green and consists of sepals. The number of Sepals generally varies from 3 to 5 in different flowers but exceptionally it may be 2 or more than 5. The protection of inner floral parts in bud condition and photosynthesis are the two main functions of calyx.

The Corolla

The second whorl of floral parts is called corolla and it consists of coloured petals. The number of petals corresponds with that of the sepals. Rarely, it is more or less than the sepals. The petals are arranged alternately with the sepals. The corolla is meant for the protection of inner floral whorls and the attraction of insects for pollination.

The Androecium

The two outer whorls of floral parts do not actually take part in the process of sexual reproduction and hence are termed as non - essential whorls. The third and the fourth whorl, on the contrary, take part in the process of sexual reproduction and hence are termed the essential whorls of the flower.

The third whorl is called androecium and consists of stamens. The number of stamens in a whorl varies from flower to flower. The stamen is the male organ of a flower. It consists of a long and slender filament and an anther at its apex. The anther usually consists of the two anther lobes joined by a connective. Each anther lobe contains numerous minute grains, the pollen grains. Each pollen grain consists of two to three cells and is called the male gametophyte. Each pollen grain produces two male gametes or male reproductive cells.

The Gynoecium

The innermost whorl of floral parts is the gynoecium. This is the female organ of a flower. The gynoecium consists of one to many carpels. Each carpel has a swollen basal portion, the ovary, which contains one to many ovules or female gametophyte. The ovary continues into a slender tube up-wards, known as style which terminates into a stigma. The stigma is the receptive organ of carpel and it receives pollen grains during pollination. The ovules are attached towards inner side of the ovary. The place where the ovules are attached in the ovary is called placenta. The placenta supplies nourishment to the developing ovule. Each ovule contains an embryo-sac, into which, there is an egg-cell or female reproductive cell (Fig. 38).

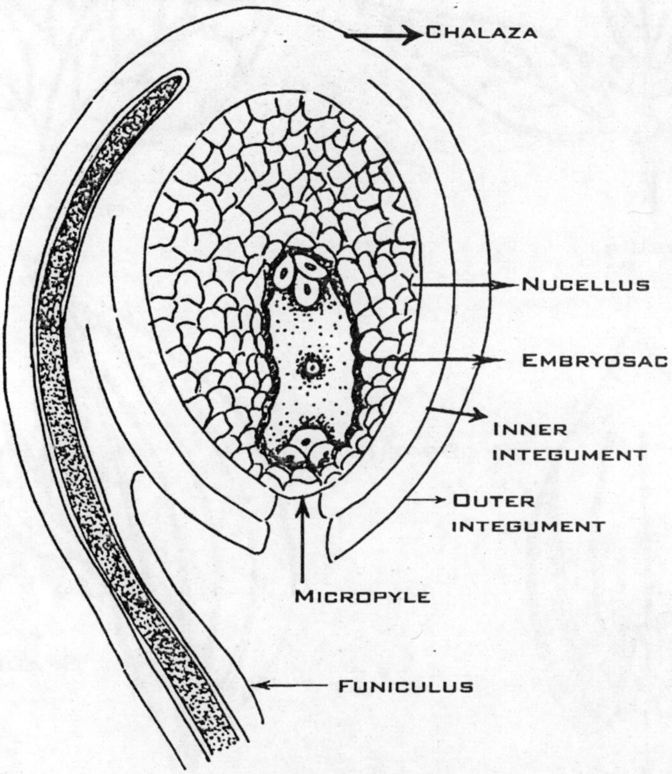

Fig. 38. Ovule

The ovary develops into a fruit and the ovules mature into seeds after fusion of male and female gametes.

Now we shall consider more details about these floral parts.

The Bract

Usually, the bracts are small and green but in certain flowers such as *Bougain- villea*, they are large and coloured. They appear like petals and hence are called petaloid bracts. The bracts of certain inflorescences are also large and brightly colored. The flowers in these inflorescences are minute and inconspicuous. Therefore, the coloured bracts perform the function of attraction of insects and bracts of inflorescences are found in *Euphorbia pulcherrima,* banana, coconut, most of the aroids and a few other plants. These are usually termed as spathes or spathaceous bracts (Fig. 39).

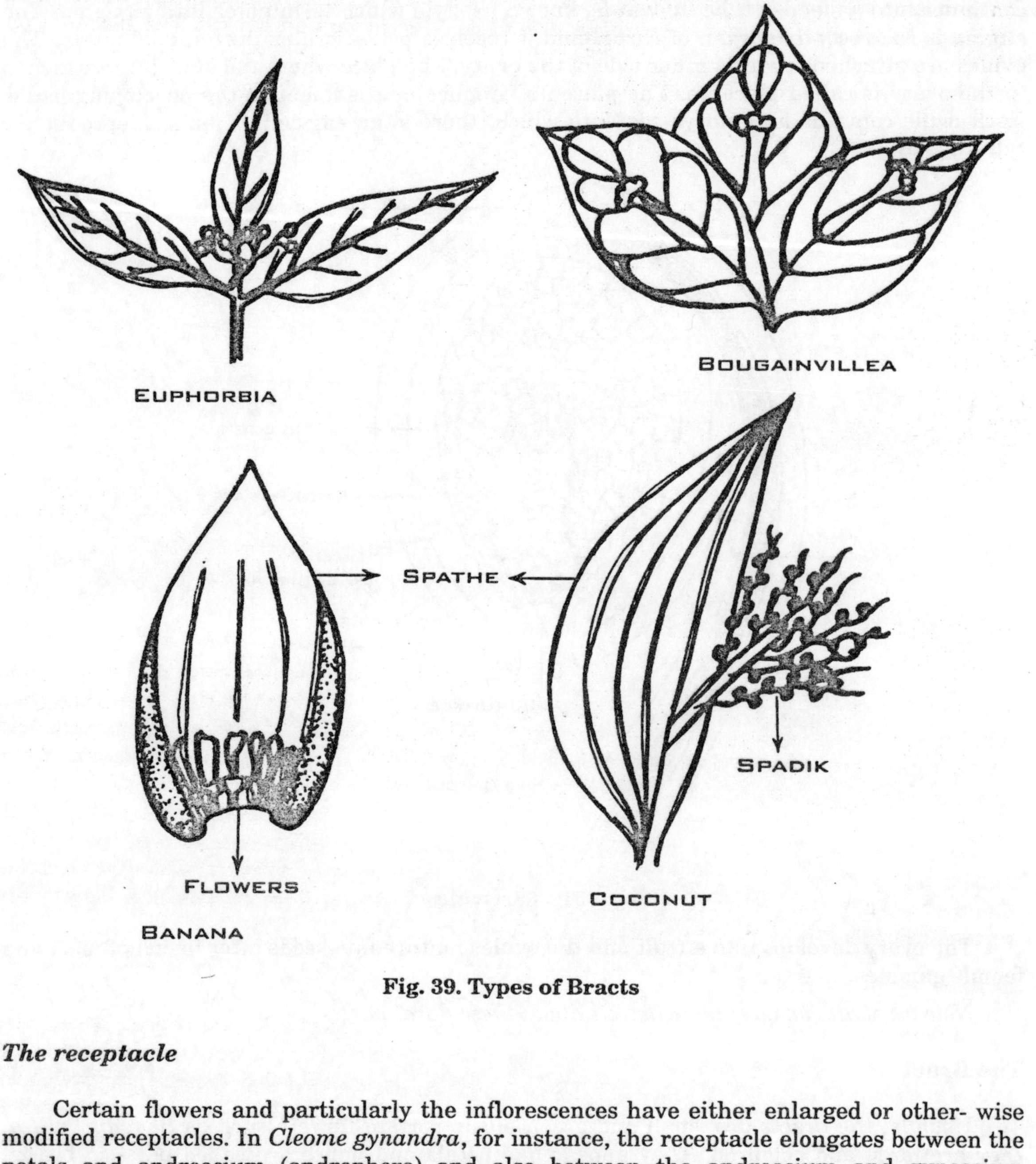

Fig. 39. Types of Bracts

The receptacle

Certain flowers and particularly the inflorescences have either enlarged or other- wise modified receptacles. In *Cleome gynandra*, for instance, the receptacle elongates between the petals and androecium (androphore) and also between the androecium and gynoecium (gynophore) (Fig. 40). The receptacle in the flower of rose is cup-like while in the flower of *Michelia champaca* it is much elongated. In *Semecarpus anacardium* and *Anacardium occidentale*, the receptacle is normal in the flower but it becomes enlarged and fleshy in fruit. The receptacles in certain plants such as apple, enclose the ovary and become fleshy in fruit. The receptacle in the head-like inflorescence of the sunflower family Asteraceae is usually flat and disc-like or less commonly it is cup-like, dome shaped, globose and so on. The receptacle of *Quercus* is capular whereas it is vessel-like in fig and banyan like plants (Fig. 41).

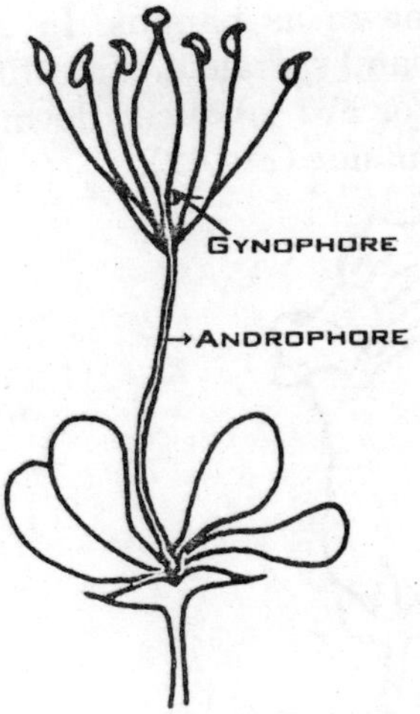

Fig. 40. Gynandrophore of Cleome

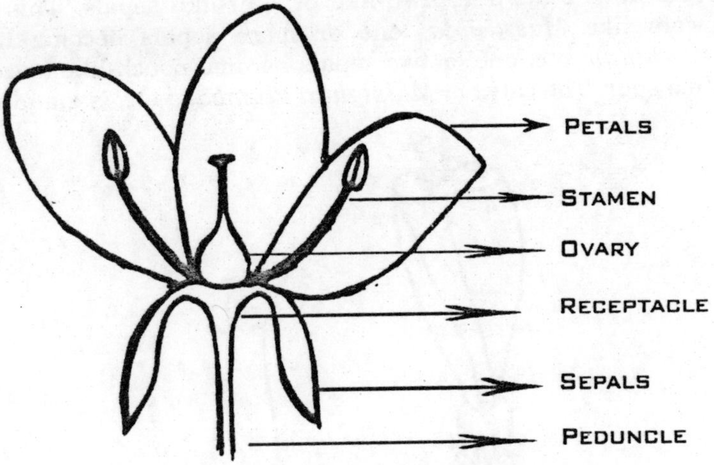

Fig. 41. Parts of a typical Flower

Complete and incomplete flower

If the flower has all the four whorls, namely calyx, corolla, androecium and gynoecium, then the flower is called as complete or perfect. If one or two of these whorls are lacking, then the flower is termed as incomplete or imperfect. If the flower has both androecium and gynoecium, then it is called as a bisexual or hermaphrodite flower, if on the contrary, one of them is lacking then the flower is called unisexual. If only the androecium is present, then it is a male flower and when only the gynoecium is present, it is a female flower. When both kinds of unisexual flowers are present on the same plant, then plant is termed as monoecious and when the male and female flower are bloom on separate plants then the plants are called dioecious. When the plant bears unisexual as well as bisexual flowers, then the flowers are termed as polygamous.

Symmetry of the flower

When the floral parts are properly arranged on the receptacle and the members of a single whorl are similar in size and shape, the flower is described as symmetrical, regular or actinomorphic. On the contrary, if the members of the same whorl (e.g. sepals, petals etc) are not similar in size and shape, or they are not arranged symmetrically, then the flower is termed as asymmetrical or zygomorphic. The flowers of *Hibiscus, Catharanthus, Ipomoea* etc. are actinomorphic while those of Bean, Coriander, *Ocimum* etc. are zygomorphic.

The Calyx

When the sepals are separate from each other or distinct, the calyx is said to be polysepalous, as in Mustard, *Ipomoea* etc. If the sepals are united, partly or wholly, the calyx is termed as gamosepalous as in *Hibiscus, Ocimum* etc. In some flowers, the calyx is coloured as in *Gloriosa,* and is called petaloid calyx. In *Hibiscus,* cotton etc. there is an additional whorl of bracts out-side the calyx. This is often called as epicalyx or the whorl of bracteoles. In the family Asteraceae, certain members like sunflower, the central florets or the disc florets have scale-like, awn-like or hair-like sepals. This is known as pappus. In certain flowers like *Mussaenda,* one or more sepals become large and petaloid. In Mustard, *Antirrhinum* etc. one or two sepals become pouch-like (saccate) or are produced downwards into a spur. The calyx of *Malpighia, Plumbago* etc. is glandular outside (Fig.42).

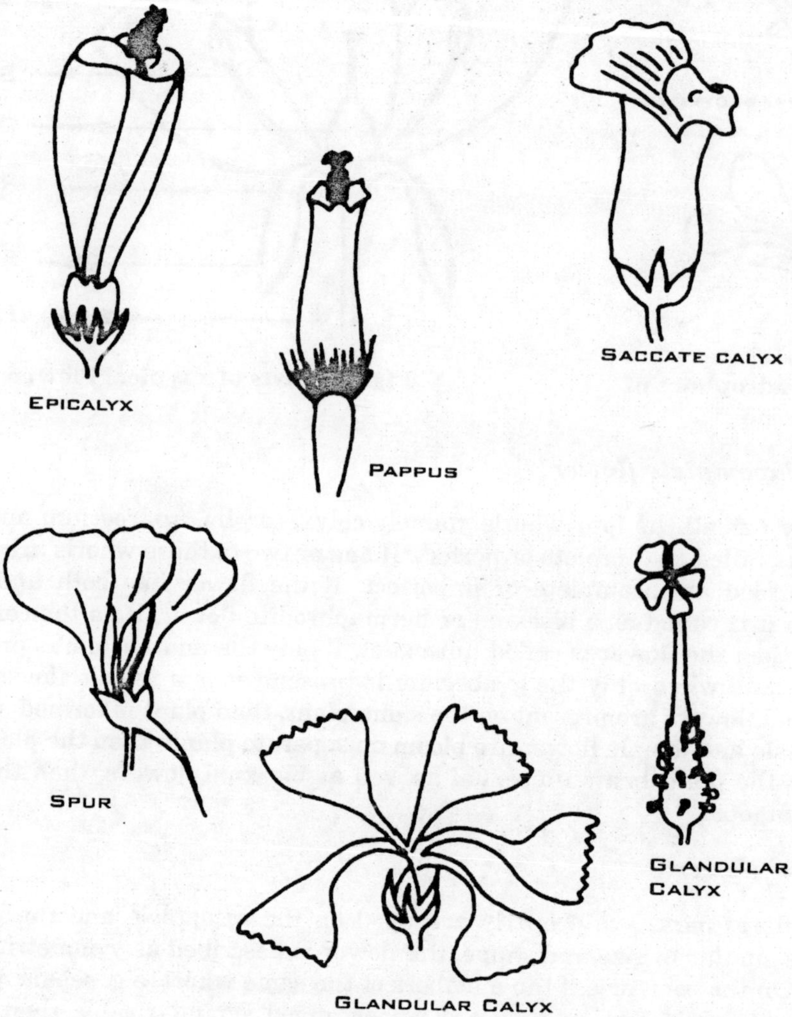

Fig. 42. Modification of Calyx

The Corolla

The constituent members of the corolla, namely, the petals may be distinct (polypetalous) or united partly or wholly (gamopetalous or sympetalous). In certain flowers the petals may be sepaloid (green). In passion flower, there is additional whorl of corolla – like members. It is called corona. Corona may develop from petals as in *Nerium* or it may develop from filaments of androecium as in *Calotropis*.

Forms of Corolla

The corolla may assume various forms due to arrangement and shape of petals. In actinomorphic, polypetalous flowers the corolla may be cruciform (cross-like) as in *Brassica*, caryophyllaceous (five petals with narrow, erect stalk and flat, spreading limb) as in pink or rosaceous, (five petals without stalks, spreading) as in rose, *Tribulus* etc. (Fig. 43). The papilionaceous corolla of bean, pea, *Butea* etc. is form of zygomorphic, polypetalous flower. This corolla appears somewhat like a butterfly (papillion). Here, there are five petals, one of which is the largest and is called standard petal or vexillum, the two, side petals, which are narrow and termed as wings or carina and the two opposite to the standard petal are more or less united along one margin and form a boat shaped structure known as the keel (Fig. 44).

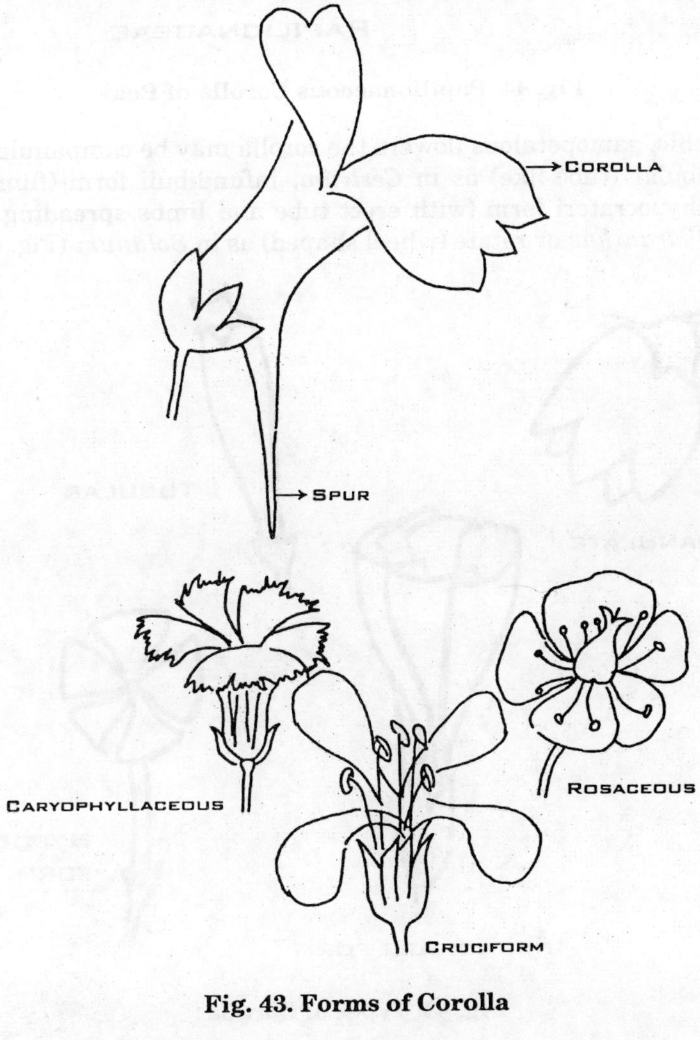

Fig. 43. Forms of Corolla

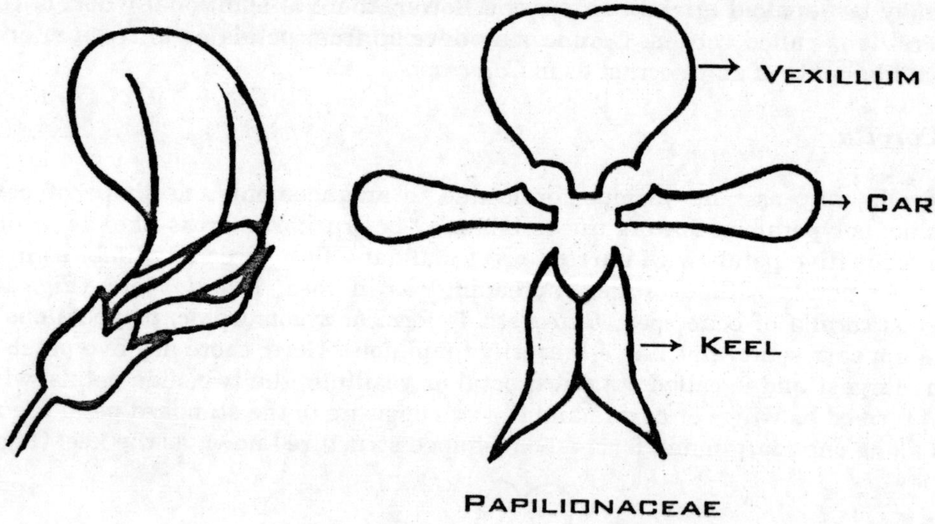

Fig. 44. Papilionaceous Corolla of Pea

In actinomorphic, gamopetalous flowers the corolla may be campanulate (bell shaped) as in *Campanula*, tubular (tube-like) as in *Cestrum,* infundibuli form (funnel shaped) as in *Ipomoea, Datura,* hypocrateri form (with erect tube and limbs spreading at right angles to the tube) as in *Catharanthus* or rotate (wheel shaped) as in *Solanum* (Fig. 45).

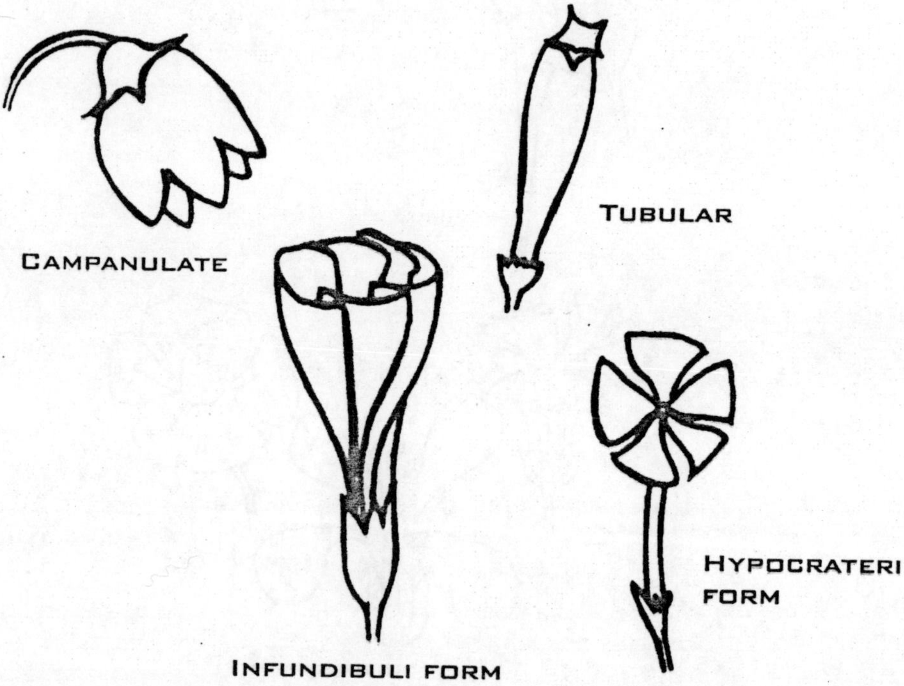

Fig. 45. Types of Corolla

In zygomorphic, gamopetalous flowers, the corolla may be ligulate (strap-shaped or tongue-shaped) as in marginal florets of sunflower or bilabiate (two lipped) as in *Antrrhinum or Ocimum* (Fig. 46).

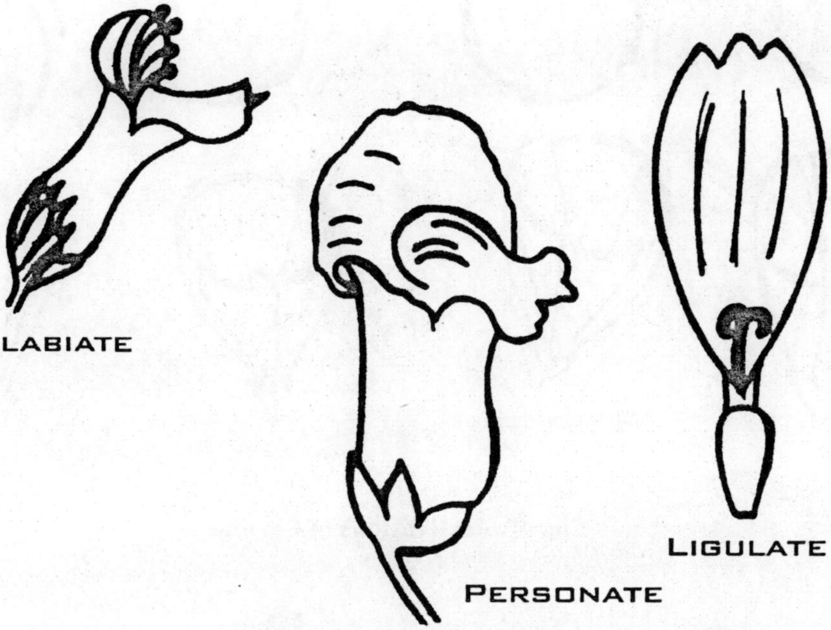

Fig. 46. Types of Corolla

Aestivation

The arrangement of sepals or petals in bud condition is called aestivation. It is valvate when members are merely touching each other by their margins, contorted when one of their margins is overlapping and the other is being overlapped, imbricate when one of the members has both the margins outside the whorl, one member has both the margins inside the whorl and other members have one margin out and the other inside the whorl. In vexillary aestivation, the largest petal envelops other petals in such a way that its both margins lie outside, both margins of the keel petals lie inside and the wing petals have one margin outside and one margin inside the whorl (Fig. 47).

The Androecium

We have already noted that each stamen has a filament and an anther. The filament may be long or short, the anther usually consists two anther lobes (two celled anther) but in the family Malvaceae, the anthers are one lobed or one celled.

The filament may be variously attached to the anther. It may be basifixed (attached at the base), dorsifixed (attached at the apex), versatile (attached in the centre) or adnate (attached throughout the length of the anther). The examples of basifixed anthers may be found in *Datura,* dorsifixed in *Bignonia*, versatile in grasses and adnate in *Michelia* (Fig. 48).

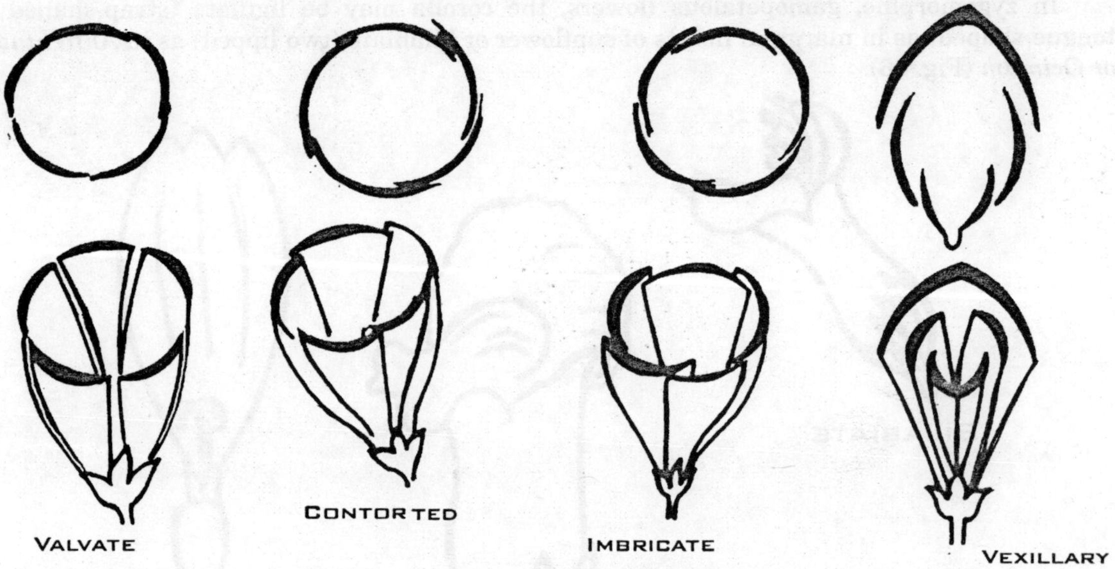

Fig. 47. Aestivations of Corolla

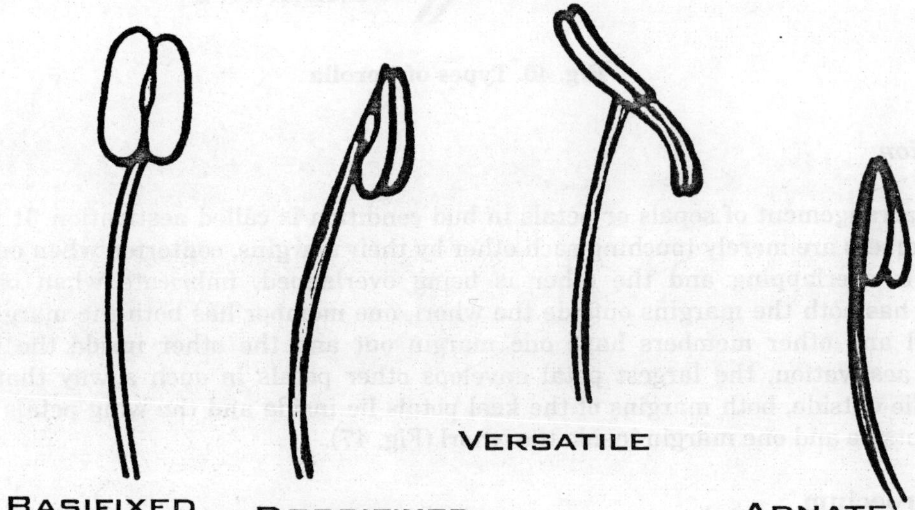

Fig. 48. Attachment of Filament with the Anther

Cohesion of stamens

The stamens may be free from each other or may be variously united. In adelphous condition, only the filaments are united, in syngenesious condition, the anthers are united and in synandrous condition, both anthers as well as filaments are united.

In the flowers of *Hibiscus*, cotton etc. The filaments are united into a single bundle (monadelphous), in pea, gram etc. the filaments are united to form two bundles (diadelphous), in *Bombax*, they are united into five or six bundles (pentadelphous or polyadelphous) (Fig. 49).

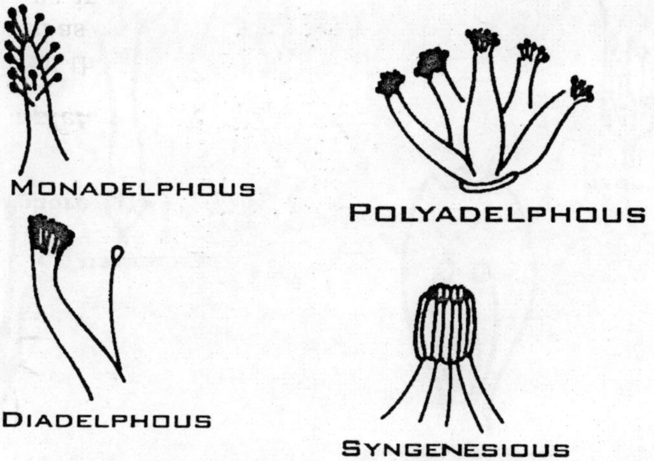

Fig. 49. Adelphy Conditions

The syngenesious condition is seen in most of the members of Asteraceae family and the synandrous condition in Cucurbitaceae family.

Adhesion of stamens

When the stamens are united with the petals by their filaments, they are described as epipetalous as in *Datura,* when united with the perianth epiphyllous as in lily, and when united with the gynoecium gynandrous, as in orchids.

Length of stamens

The filaments of stamens in a flower may be of equal length or they may be of unequal length as in *Ipomoea*. The stamens may be coming out of the corolla-tube (exserted) or included within the tube. In the family Brassicaceae, there are usually six stamens in a flower, out of these, four are longer than the other two (tetradynamous condition). In the zygomorphic and gamopetalous flowers, there are usually four stamens, two of which are long and two are short (didynamous) (Fig. 50).

Dehiscence of anthers

The anther lobes break-open when the pollen grains included in them become mature. This is called the dehiscence of anthers. The anther lobes may dehiscence lengthwise (longitudinal), crosswise (transverse), by means of pores (porous) or by means of valves (valvular). The longitudinal dehiscence may be seen in *Datura*, transverse in *Ocimum*, porous in Solanum and valvular in *Cinnamomum* (Fig. 51).

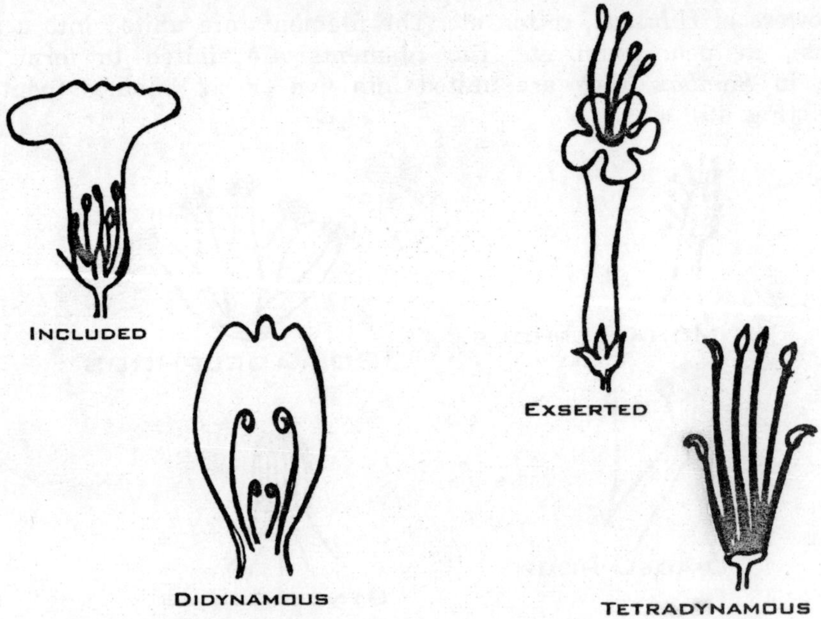

Fig. 50. Relative lengths of Stamens

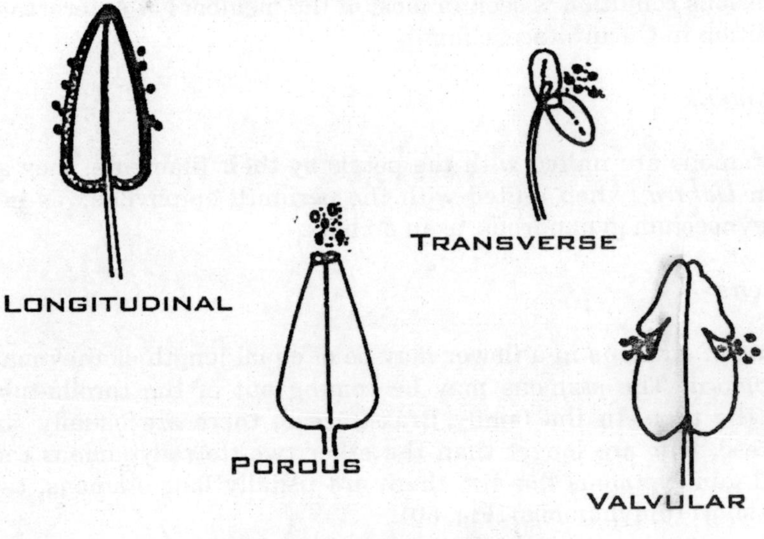

Fig. 51. Dehiscence of Anthers

The Gynoecium

If the gynoecium is composed of one carpel (monocarpellary) as in bean, pea etc. then it is described as simple. The compound gynoecium is formed of more than one carpel (multicarpellary). In the compound gynoecium, if the carpels are united then it is called syncarpous gynoecium and if the carpels are distinct, it is apocarpous one (Fig. 52).

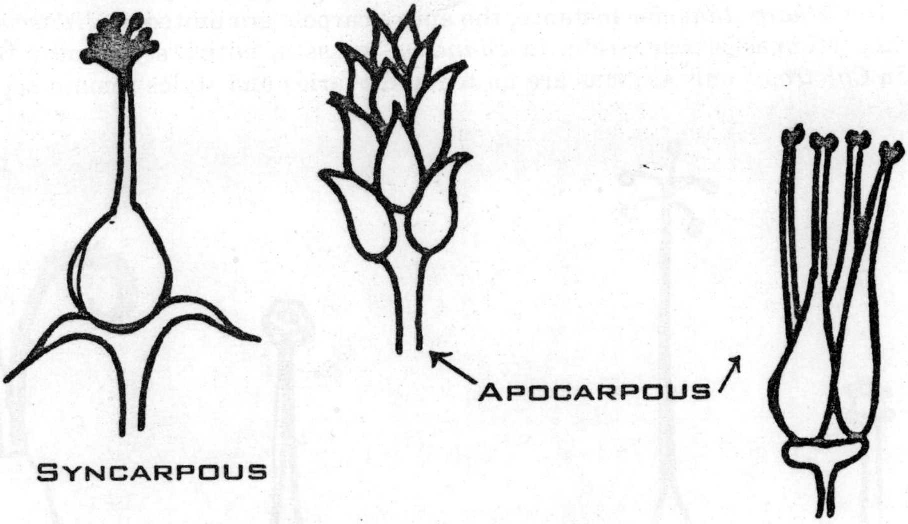

Fig. 52. Types of Carpels

The style may be terminal or at the apex of the ovary or it may occupy a lateral position as in Morus. If the style is arising from the base of the ovary, as in *Ocimum,* then it is called as gynobasic (Fig. 53).

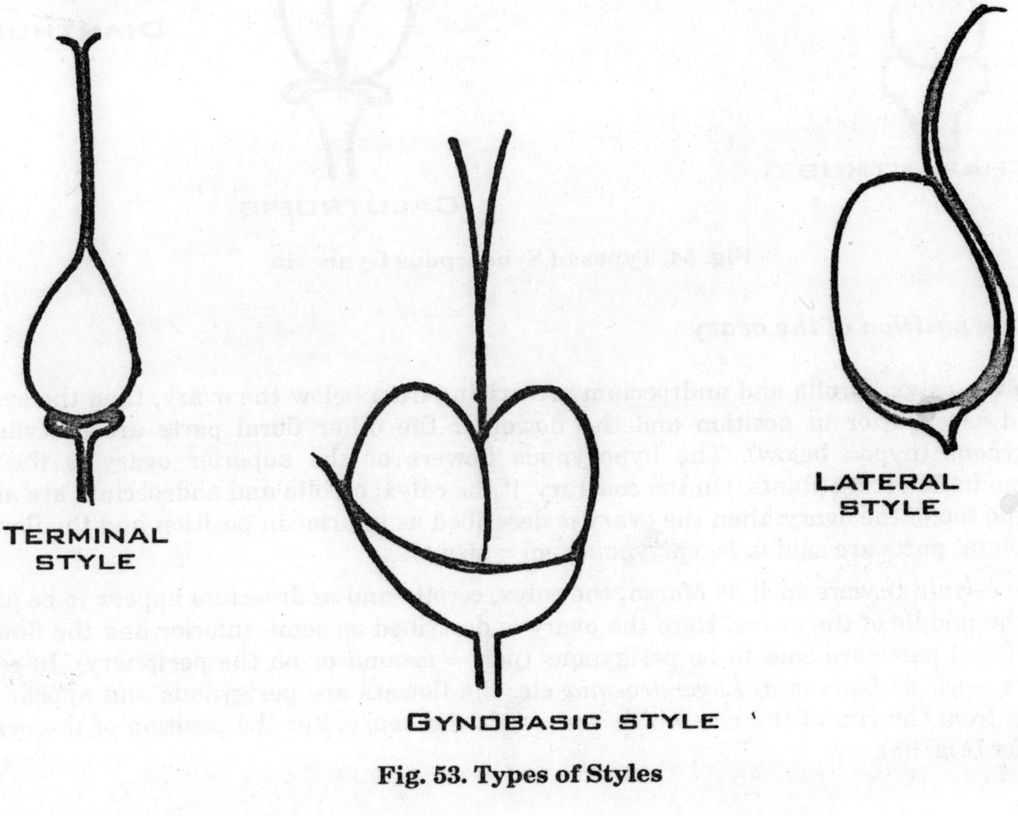

Fig. 53. Types of Styles

In the compound, syncarpous gynoecium, the entire carpels may be united or their part may be so. In *Catharanthus*, for instance, the entire carpels are united, in *Hibiscus*, the style branches and stigmas are separate, in *Dianthus*, grasses, entire styles may be separate where as in *Calotropis* only stigmas are united and ovaries and styles remain separate (Fig. 54).

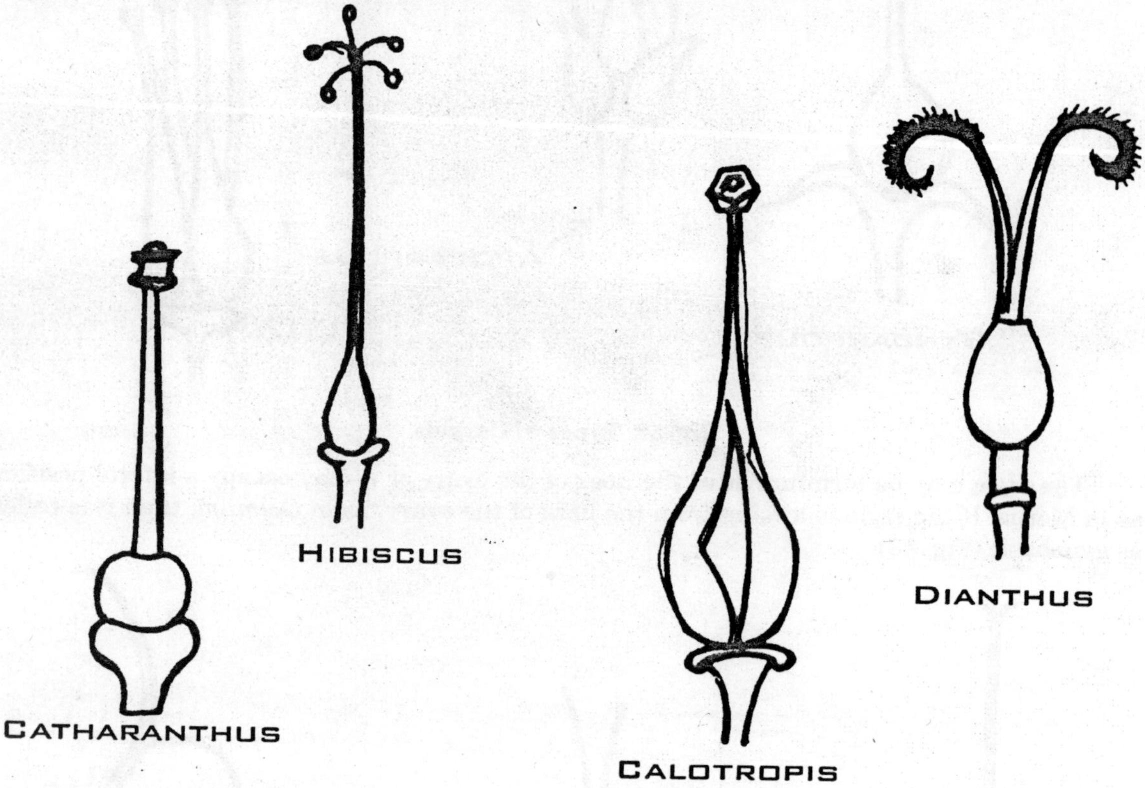

Fig. 54. Types of Syncarpous Gynoecia

Relative position of the ovary

If the calyx, corolla and androecium are arising from below the ovary, then the ovary is termed as superior in position and the flower or the other floral parts are described as hypogynous (hypo= below). The hypogynous flowers or the superior ovary is the most common in flowering plants. On the contrary, if the calyx, corolla and androecium are arising from the top of the ovary, then the ovary is described as inferior in position and the flower or other floral parts are said to be epigynous (epi = above).

In certain flowers such as *Maesa*, the calyx, corolla and androecium appear to be arising from the middle of the ovary. Here the ovary is described as semi- inferior and the flower or other floral parts are said to be perigynous (peri = around or on the periphery). In certain flowers, such as *Lawsonia, Lagerstroemia* etc. the flowers are perigynous and appear to be arising from the rim of the cup which surrounds the ovary, but the position of the ovary is superior (Fig. 55).

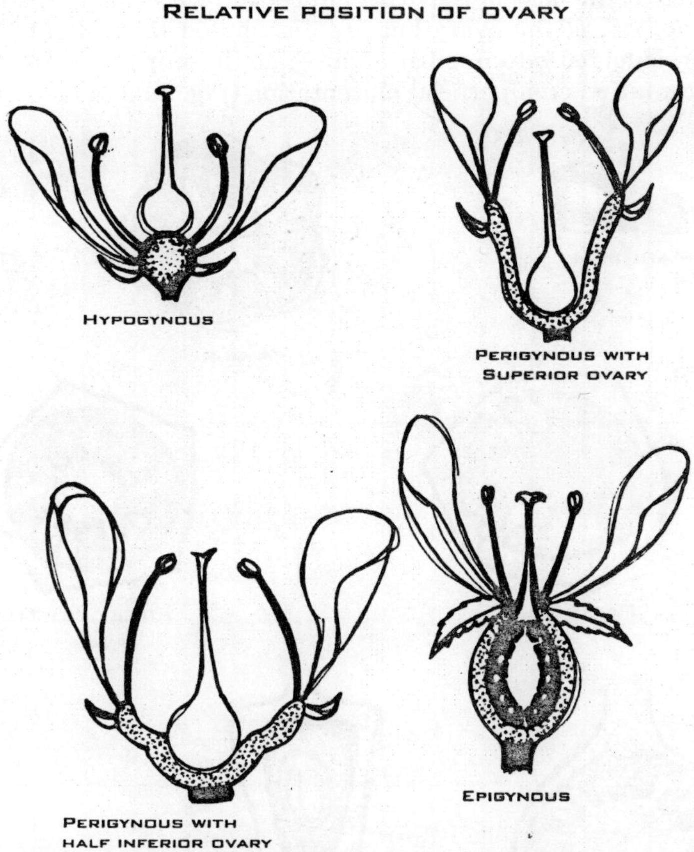

Fig. 55. Position of Ovary

The placentation

The arrangement of placentas in the ovary is called as placentation. There are seven main types of placentation. The placentation in monocarpellary gynoecium of pea, bean etc. is situated along one margin of the carpel and is described as marginal placentation. In the compound gynoecium with free carpels, the placentation in the individual carpel is actually marginal, but to distinguish it from the typical marginal, placentation of the monocarpellary gynoecium, it is described as axil. In the syncarpous gynoecia the placentation may be axile and ovules are attached to the central axis. The ovary, in this case, is many chambered. The axile placentation is the most common type found in syncarpous gynoecia. The placentation is described as parietal when the placentas are situated along the wall of the ovary. Here the ovary is one chambered and there are as many placentas as there are carpels. The parietal placentation is seen in mustard, *Argemone*, cucumber etc. The central placentation is seen in *Dianthus*. Here the ovary is one chambered and placentas are situated on the central axis. This axis runs from the base of the ovary and continues up-to the top. The free central placentation in *Primula, Anagallis* etc. is vey similar to central placentation but the axis is not continuous but ends much below the top of the ovary. The basal placentation, as seen in

Plumbago, is situated at the base of the ovary. In certain water plants, the placentas are not restricted to any one part of the ovary but are distributed throughout the inner wall. The inner wall, however, is folded into number of lamellae (incomplete walls). This type is called as the lamellar placentation or superficial placentation (Fig. 56).

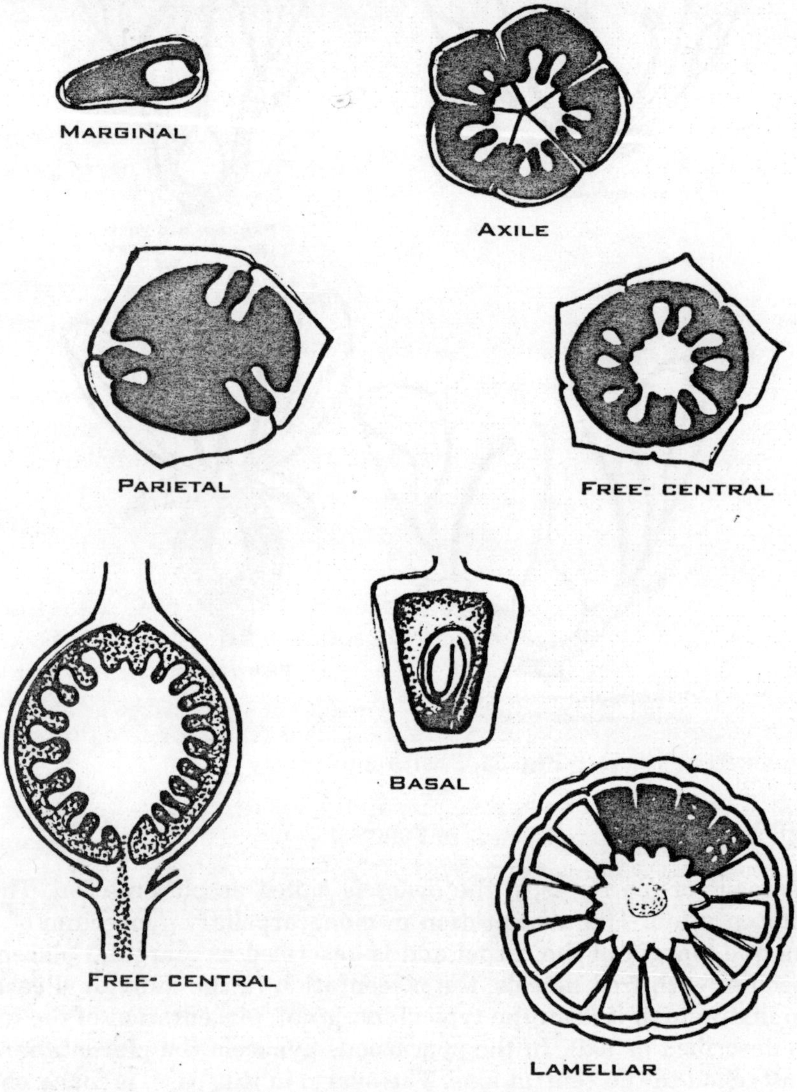

Fig. 56. Types of Placentation

The Inflorescence

The flowers of *Michelia, Datura* etc. are large and are borne in the axils of leaves or at the end of the branches, they are usually solitary. But those of *Ocimum, Jasminum* etc. are relatively small and are borne in a bunch. Such bunches of flowers borne by the plants are termed inflorescences. Many small flowers in an inflorescence thus become conspicuous.

There are two main types of inflorescences, namely, the racemose or indeterminate and the cymose or determinate types.

In the racemose types of inflorescences, the main axis bears flowers on the lateral sides and the terminal bud continues to grow more or less continuously. Here the lowest flowers are the oldest and the uppermost ones are the youngest. This sequence of maturation of flowers in an inflorescence from the base to the apex is called acropetal succession or centripetal sequence. On the contrary, in cymose types of inflorescences, the terminal bud itself gives rise to a flower and further growth of the main axis is stopped. The later flowers develop from the lateral buds below the first flower and this sequence is maintained in the development of future flowers. Here the maturation sequence of flowers is from apex to the base or basipetal or centrifugal.

The types of Racemose Inflorescence

1. The raceme — In *Caesalpinia*, Mustard etc. there is a single main axis which bears flowers on pedicels. The lowest flower is the oldest.

2. The panicle — When the main axis of the raceme gives out branches and these secondary or tertiary branches bear flowers in an acropetal sequence, then the inflorescence is called as panicle. Mango, Neem, *Moringa* are some examples of panicle inflorescence.

3. The spike — In this type, the main axis is long and bears sessile flowers Ex. *Achyranthes, Adhatoda, Amaranthus* etc.

4. The spikelet — The flowers of grasses, sedges etc. are very minute and are borne in small spikes called spikelets. Here the minute flowers or florets are subtended by dry, chaffy bracts. In a grass spikelet, there are two empty glumes at the base, one above the other and facing each other. They are called lower glume and the upper glume. The third and the subsequent bracts are called lemmas (singular-lemma). They subtend one floret each in their axil. The lemmas usually have one more bract opposite them. It is called palea. Thus, the lemma and the palea enclose a floret.

5. The catkin — This is a spike bearing unisexual flowers as in *Morus*, *Acalypha* etc. An ament is a kind of catkin, usually bearing male and female flowers (androgynous) and the inflorescence falls off as a whole after maturity. The ament inflorescence is seen in *Quercus, Alnus, Betula* etc. The typical catkin is usually hanging or pendulous whereas the ament may be pendulous or erect.

6. The spadix — In this type, the main axis is thick and fleshy and the flowers are sessile and often sunk into the axis. The axis is subtended and enclosed by a large bract called spathe. The spadix inflorescence is the characteristic feature of aroids or the members of Araceae family.

If the main axis of a spadix is branched as in coconut and other palms, it is called as the compound spadix.

7. The corymb — In this type, the main axis is rather short and flowers appear to be arranged in a straight line. Ex. *Quisqualis, Lantana, Calotropis* etc.

8. The umbel — Here the main axis is extremely short simulating a single point and flowers are borne like a shape of an umbrella on short or long pedicels. Ex. *Centella.* The compound umbel is a branched umbel as in coriander.

9. The capitulum — This type of inflorescence is seen in Sunflower and other members of the Asteraceae family. Here the receptacle or main axis is flat, concave or convex and is subtended by few to many bracts in whorl (involucral bracts). The flowers are borne in such a way that the peripheral or marginal flowers are older and central flowers are younger. All florets (small flowers) in a capitulum may be similar as in *Vernonia, Dahlia* etc. or the peripheral or marginal may be different from the central ones as in Sunflower. The individual florets may be bracteate (paleaceous) or ebracteate (not paleaceous) (Fig. 57F).

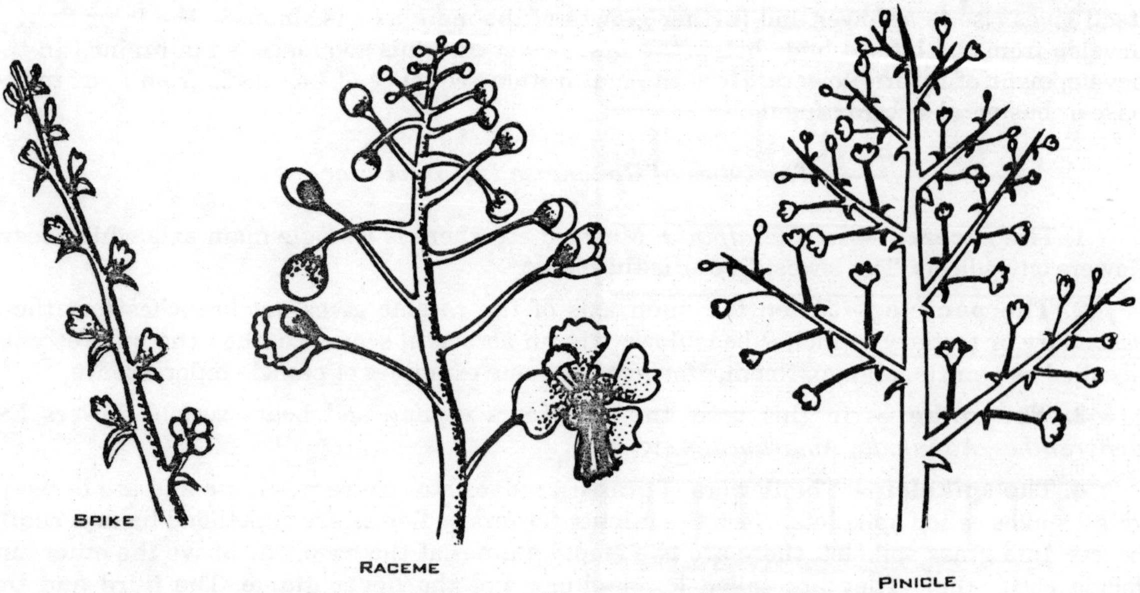

Fig. 57A. Types of Inflorescence

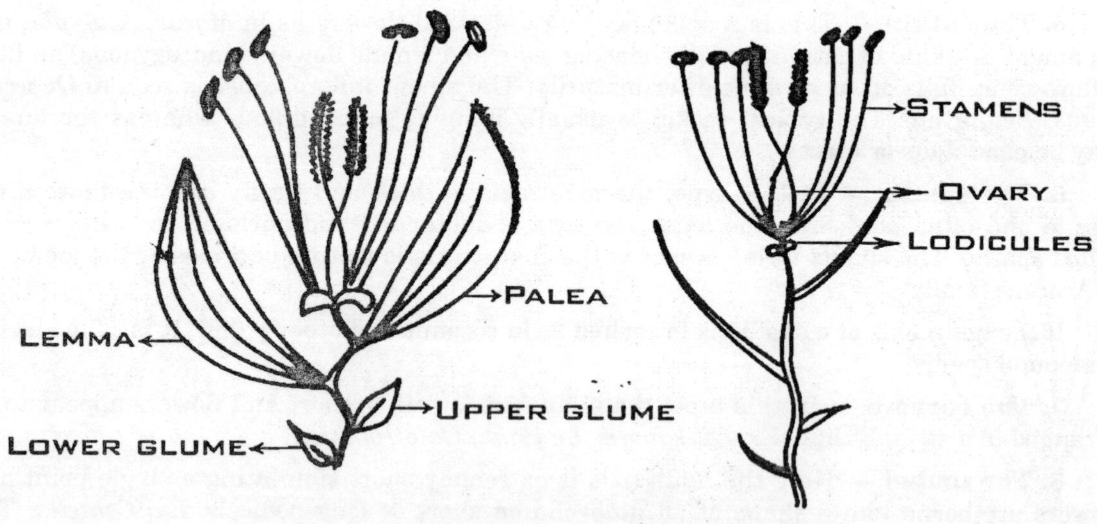

Fig. 57B. Spikelet of Grass

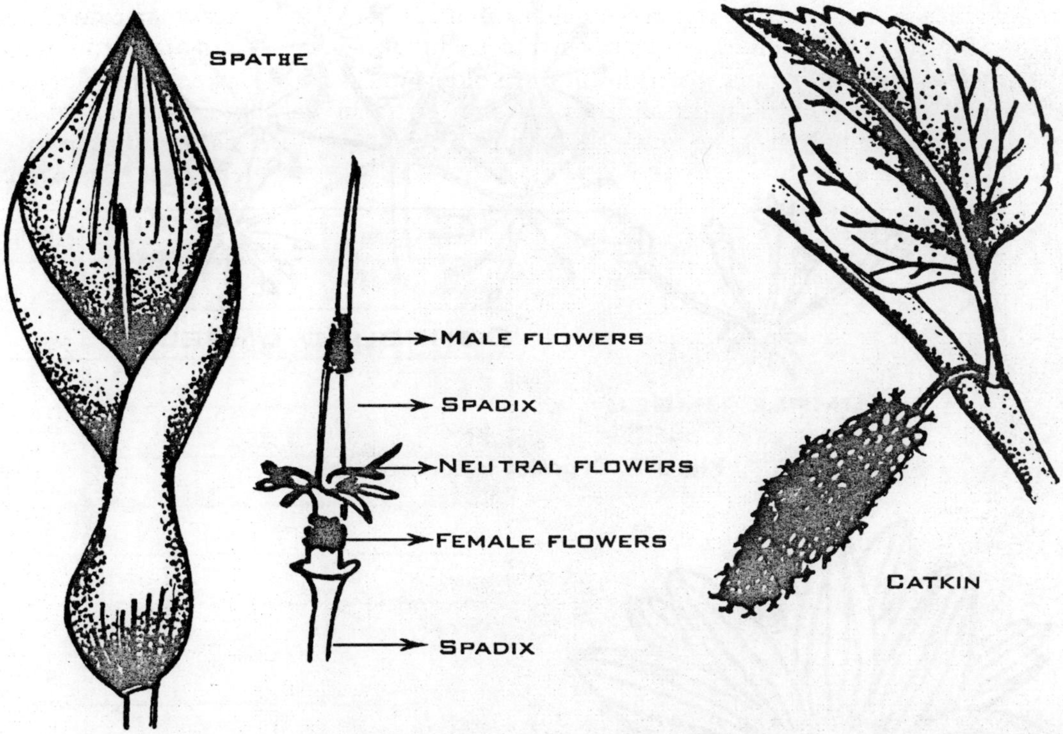

Fig. 57C. Types of Inflorescence

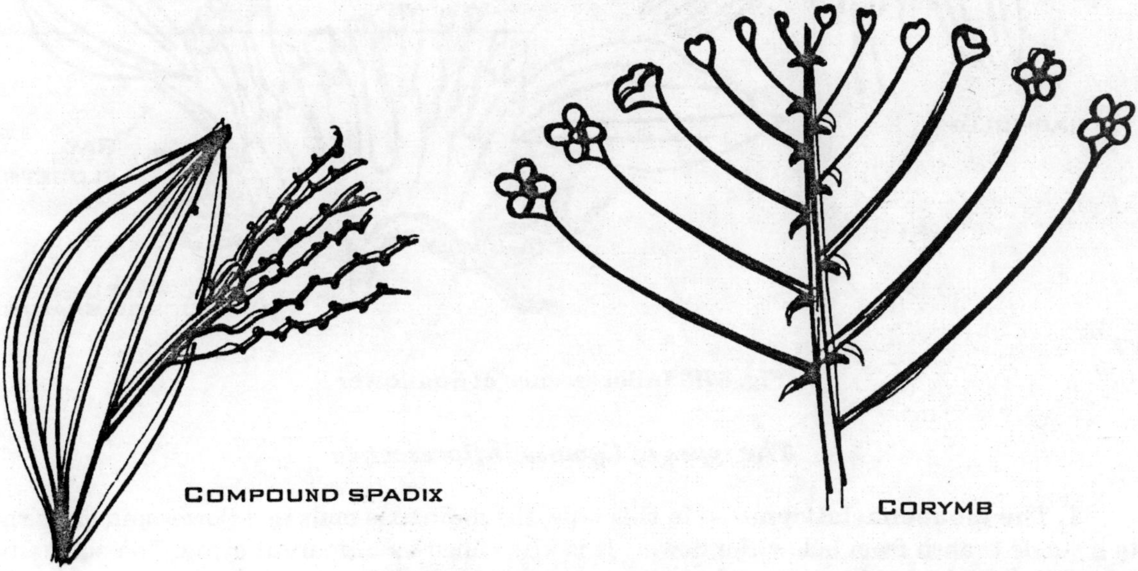

Fig. 57D. Types of Inflorescence

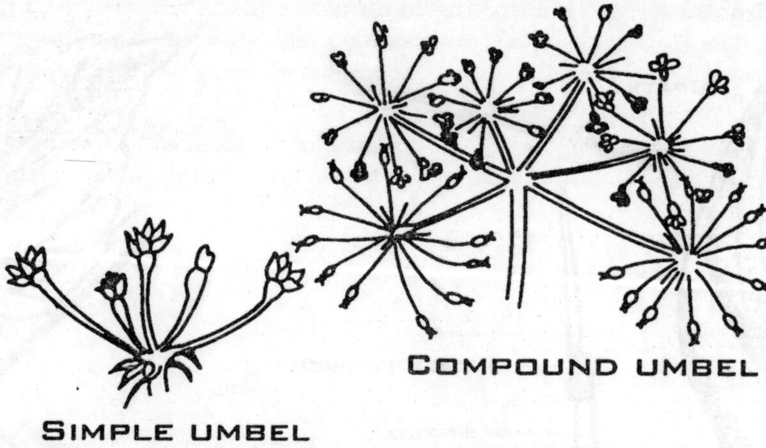

Fig. 57E. Types of Inflorescence

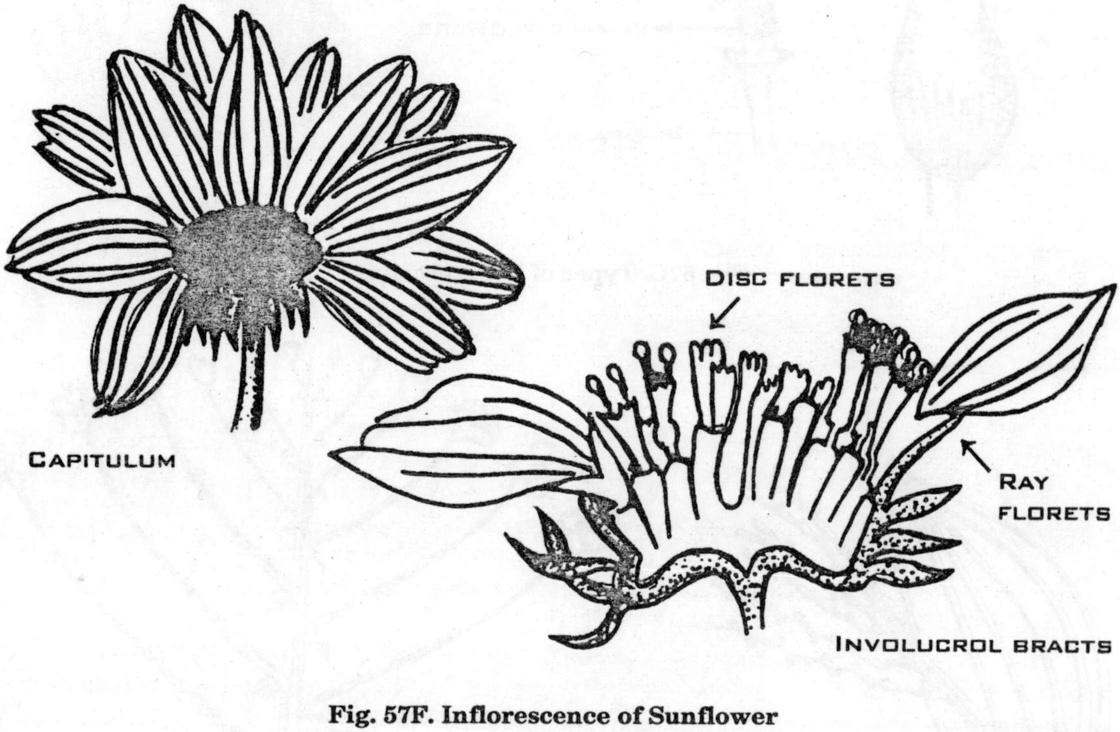

Fig. 57F. Inflorescence of Sunflower

The types of Cymose Inflorescence

1. The monochasial cyme — In this type, the main axis ends in a flower and gives rise to a single branch from below the flower. It is also called as uniparous cyme. The successive branches may be on the same side of the main axis (helicoid cyme) as in *Begonia, Hamelia* etc. or alternately on opposite sides (scorpioid cyme) as in *Heliotropium, Ehretia* etc.

2. The dichasial cyme — Here the main axis ends in a flower and gives rise to two branches from below the flower. This process may end-up with three flowers or may repeat and produce several flowers in the cyme. Examples of dichasial or biparous cymes may be seen in Jasmine, *Nyctanthes* etc.

3. The polychasial cyme — In this type the main axis, after ending in a flower, gives rise to three or more branches from below the flower as in *Nerium, Plumeria* etc. This is also called as multiparous cyme (Fig. 58).

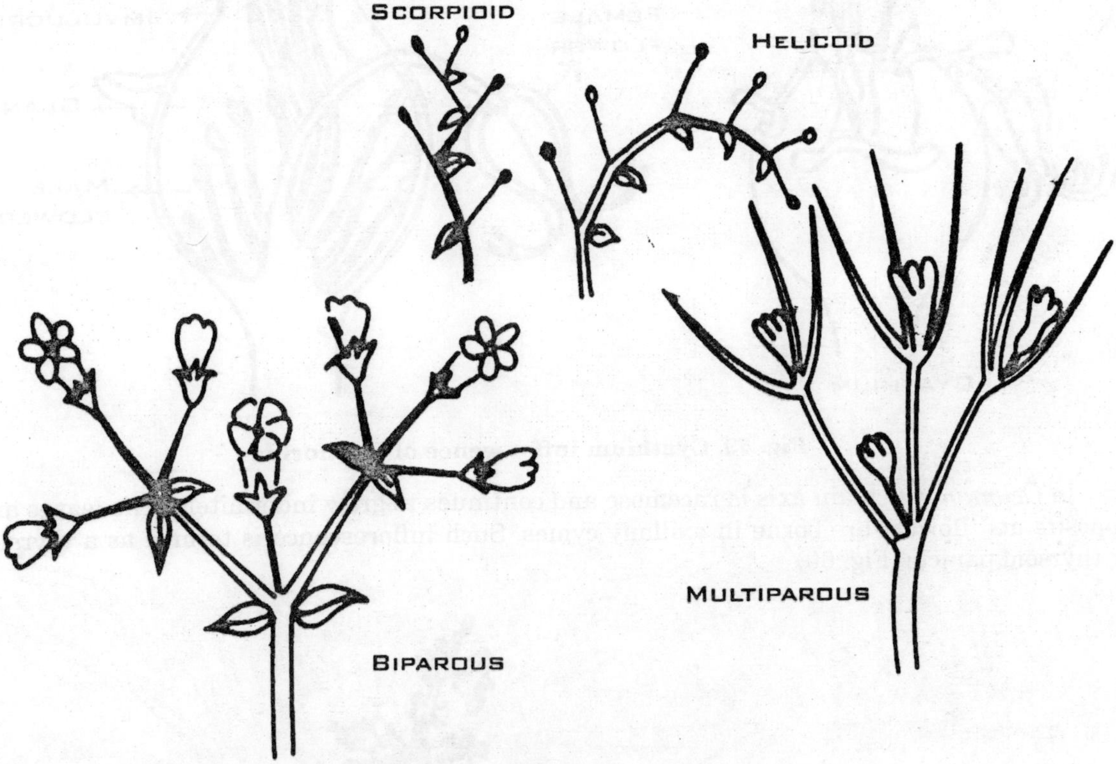

Fig. 58. Types of Cymose Inflorescence

The compound or mixed types of Inflorescences

The main two types of inflorescences described above, are the common ones. But in addition to these, there are some mixed or compound types where the main axis is racemose while the branches are cymose. There are also some special types which cannot be fitted into above two types.

In *Euphorbia*, for instance, the inflorescence is called a cyathium. Here the involucral bracts form a cup-like structure. This cup is provided with a large nectary and encloses many male flowers surrounding a single female. These flowers are without calyx and corolla (naked flowers). The males are represented by a single stamen and the females by a single gynoecium. All flowers have a small peduncle on which they are jointed (Fig. 59).

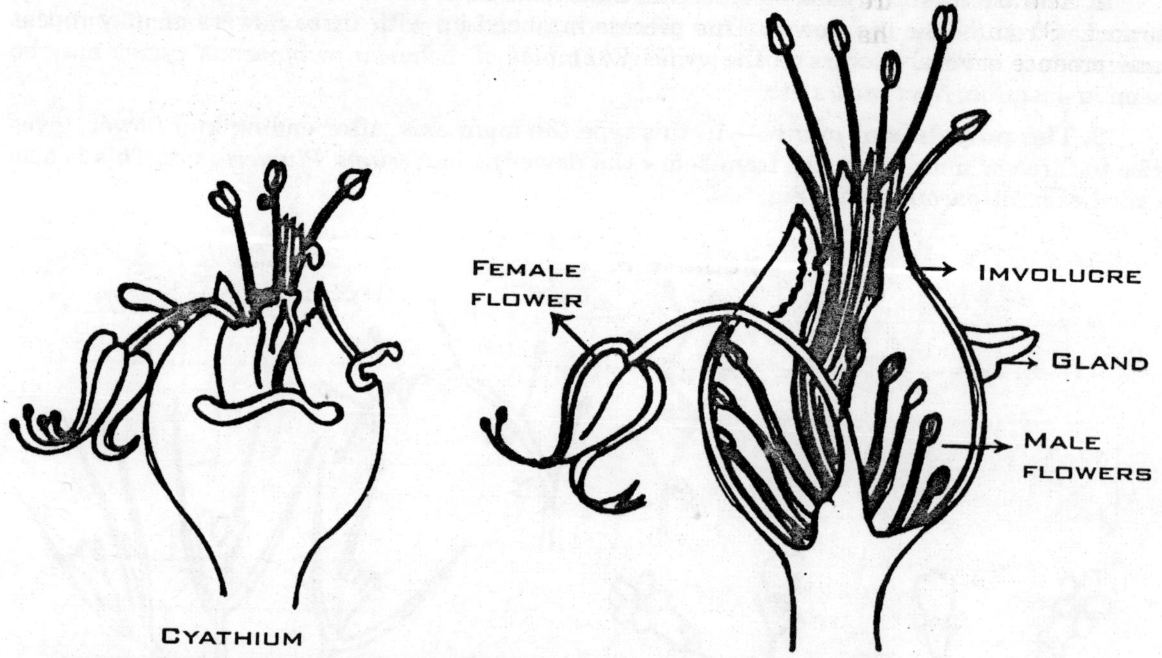

Fig. 59. Cyathium Inflorscence of Euphorbia

In *Ocimum*, the main axis is racemose and continues to grow indefinitely. The leaves are opposite and flowers are borne in axillany cymes. Such inflorescence is termed as a thyrsus or thyrsoid panicle (Fig. 60).

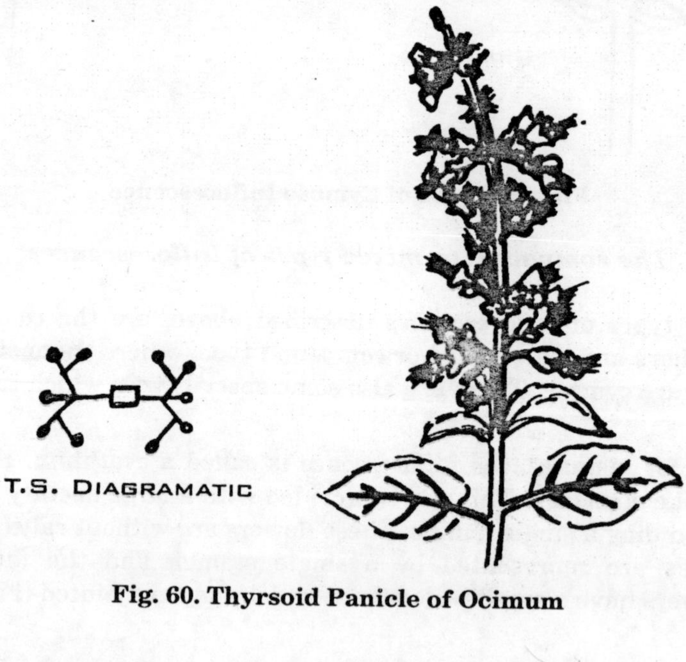

Fig. 60. Thyrsoid Panicle of Ocimum

In *Leucas*, the opposite leaves subtend cymes but the flowers are sessile and densely packed. So that they appear to be whorled. This whorl is called as verticil and the inflorescence is termed as verticillaster (Fig. 61).

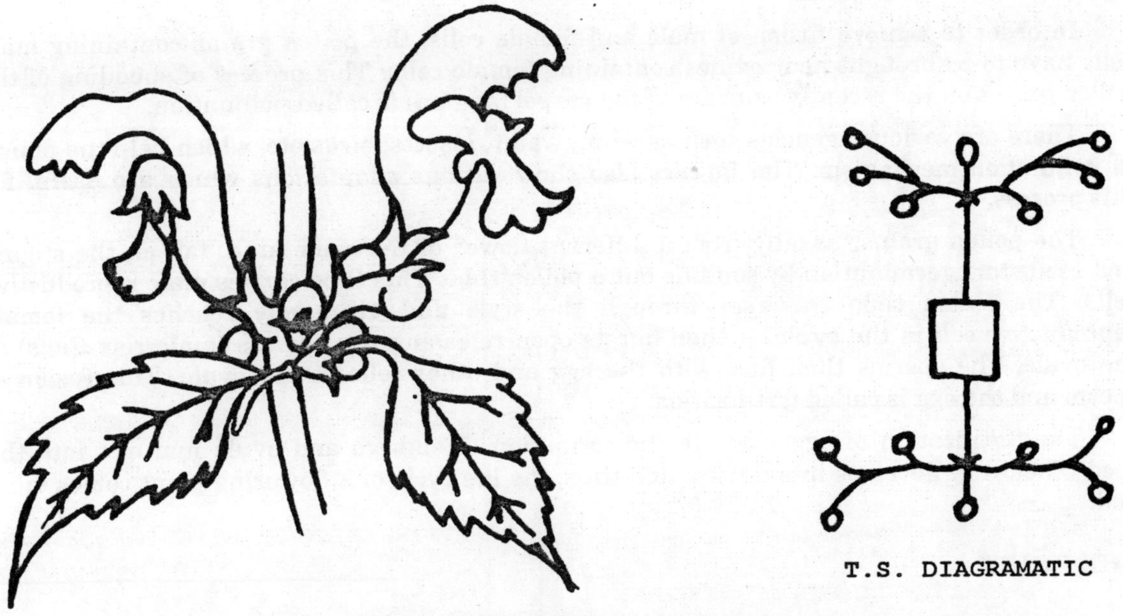

Fig. 61. Verticillaster of Leucas

In *Ficus*, the main axis forms a closed vessel leaving a small opening at the distal end. This closed vessel includes minute, unisexual and neutral flowers. This inflore-scence is called as hypanthodium and is characteristic of all species of *Ficus* (Fig. 62).

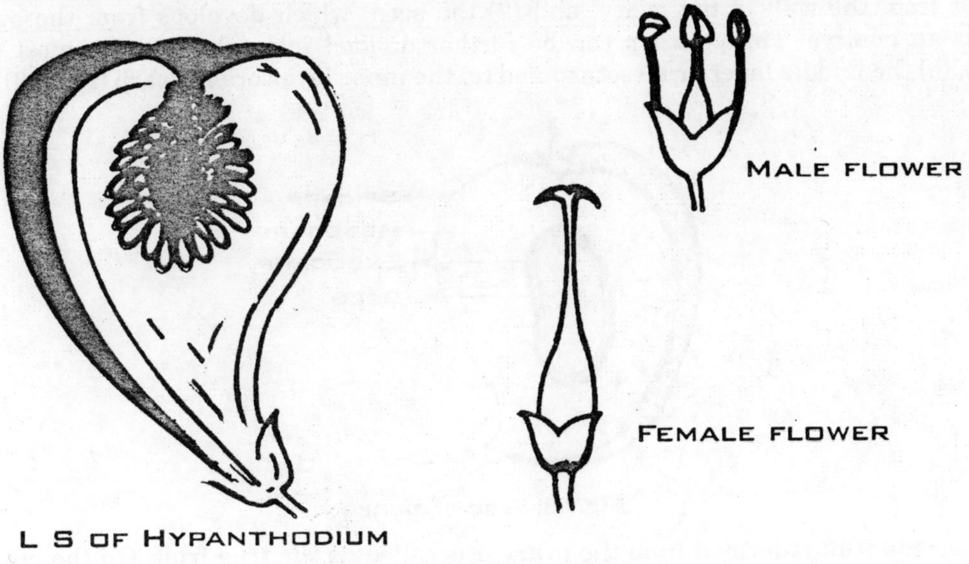

Fig. 62. Inflorescence of Ficus

The Pollination

The male and female organs of the flower, namely, stamens and carpels, produce male and female reproductive cells.

In order to achieve fusion of male and female cells, the pollen grains containing male cells have to be brought near ovules containing female cells. This process of shedding of the pollen grains on the receptive surface of the carpel (stigma) is called pollination.

There are various agencies such as wind, water, insects, birds etc. which help the plants in pollination mechanism. The flowers also show various adaptations which are useful for this process.

The pollen grains, usually from a different flower of the same kind, fall on the stigma and soon start germination by sending out a pollen tube. This tube carries male reproductive cells. The pollen tube traverses through the style and ultimately reaches the female reproductive cell in the ovule. It then bursts open releasing two sperms (male rep. Cells) in the ovule. The sperms then fuse with the egg and other cells in the ovule. This fusion of sperm and the egg is called fertilization.

The fertilization of egg leads to the formation of embryo and ovule matures into the seed. The ovary develops into a fruit, and thus, the life cycle of a flowering plant comes to an end.

THE FRUIT

The fruit is an important organ of a plant which is responsible for the propagation of a plant by way of seed.

The fruits are of various kinds and often characteristic of the plants.

A typical fruit consists of (1) an outer covering or fruit wall called as pericarp. This develops from the wall of the ovary, and (2) the seed, which develops from the ovule and contains an embryo. The pericarp can be further divided into (a) the outer most cover or epicarp, (b) the middle layer or mesocarp and (c) the inner layer or endocarp (Fig. 63).

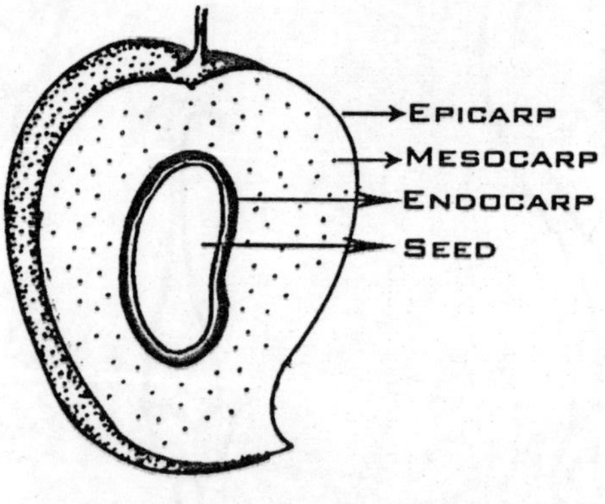

Fig. 63. Fruit of Mango

When the fruit is formed from the ovary it is called as the true fruit. On the other hand a fruit formed from any other part of the flower is called the false fruit. Mango, guava, banana etc. are the examples of true fruit where as apple, fleshy fruit in cashew nut, fig, pineapple, jackfruit etc. are examples of false fruits.

Classification of fruits

The true or false fruits are classified into three main types, namely, (1) simple, (2) aggregate and (3) composite or multiple fruits.

1. Simple fruits

A fruit formed from the simple or multicarpellary syncarpous gynoecium is called simple fruit. The simple fruit may be true or false, dry or fleshy, dehiscent or indehiscent or splitting at maturity, one seeded or many seeded.

Dry, dehiscent fruits — In this section there are four subtypes of fruits, namely (a) Legume or pod, (b) Follicle, (c) Siliqua and (d) Capsule.

The legume is a typical fruit of the members of Fabaceae, Caesalpiniaceae and Mimosaceae. It develops from the superior mono-carpellary gynoecium with marginal placentation. When mature, this fruit dehisces along both the sutures (margins) as in bean, pea etc.

The follicle also develops from the superior, monacarpellary gynoecium or polycarpellary apocarpous gyncecium. It dehisces along ventral suture only. The follicles are often seen in groups of two or more. Examples of follicles may be found in *Catharanthus, Calotropis, Michelia, Sterculia* etc.

The siliqua is narrow, long fruit developing from the superior, bicarpellary syncarpous gynoecium with parietal placentation. It dehisces on both the sutures from the base upwards. This fruit usually has a false septum or a replum in between the two placentae. The siliqua is a common fruit of mustard, raddish etc. When the siliqua is very short and has only few seeds as in *Lepidium*, it is termed as silicula (Fig. 64).

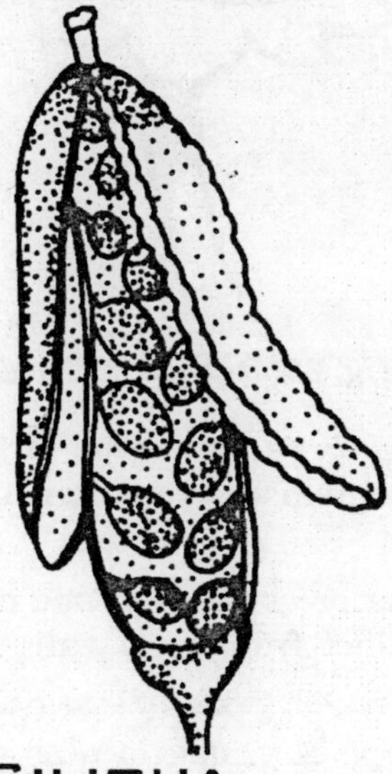

Fig. 64. Dehiscent Fruit

The capsule develops from multicarpellary syncarpous, superior or inferior gynoecium with various kinds of placentation. It dehisces in various ways and is classified further into six subtypes according to the mode of dehiscence. The capsule is termed as loculicidal when it dehisces along the back of the locules as in cotton, *Hibiscus* etc. It is termed as septicidal when it dehisces along the septa as in *Aristolochia* and septifragal when it dehisces partly leaving the central axis intact as in *Datura*. In *Portulaca, Celosia* etc. the capsule dehisces transversely and is called pyxidium or circumscissile capsule. The capsule of Papaver dehisces by means of apical pores whereas that of *Campanula* dehisces by means of basal pores. Here the dehiscence is termed as porous dehiscence (Fig. 65).

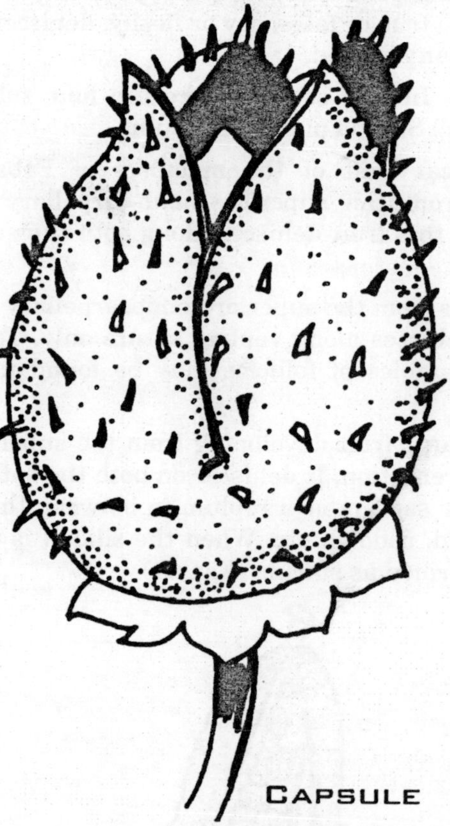

Fig. 65. Dehiscent Fruit

Dry, indehiscent or achinial fruits — This type of fruits are mostly single seeded and their entire cover is dry. They do not dehisce even after maturation. Commonly there are five kinds of these fruits.

Achene develops from the superior or inferior, one ovuled or many ovated but finally 1-seeded fruit. The fruit wall is distinct from seed wall. This kind is seen in *Chenopodium, Clematis* etc.

Caryopsis develops from bicarpellary, one locular gynoecium and possesses one seed. The seed wall is fused with fruit wall. The seed, thus cannot be separated from the fruit. Fruits of most of the grasses such as rice, wheat, maize, jowar etc. are of this kind.

Nut develops from polycarpellary, one loculed, one ovuled superior ovary. The fruit wall is thick and hard. This kind is seen in *Oak, Polygonum, Cyperus* etc. Cypsella develops from bicarpellary, inferior, one locular and one ovuled ovary.This fruit has dry, persistent calyx (pappus) on top of it. This kind of fruit is found in most of the members of sunflower family.

Samara is an achinial type of fruit in which the fruit wall develops thin papery wings as in *Ventilago, Hiptage, Terminalia arjuna* etc. (Fig. 66).

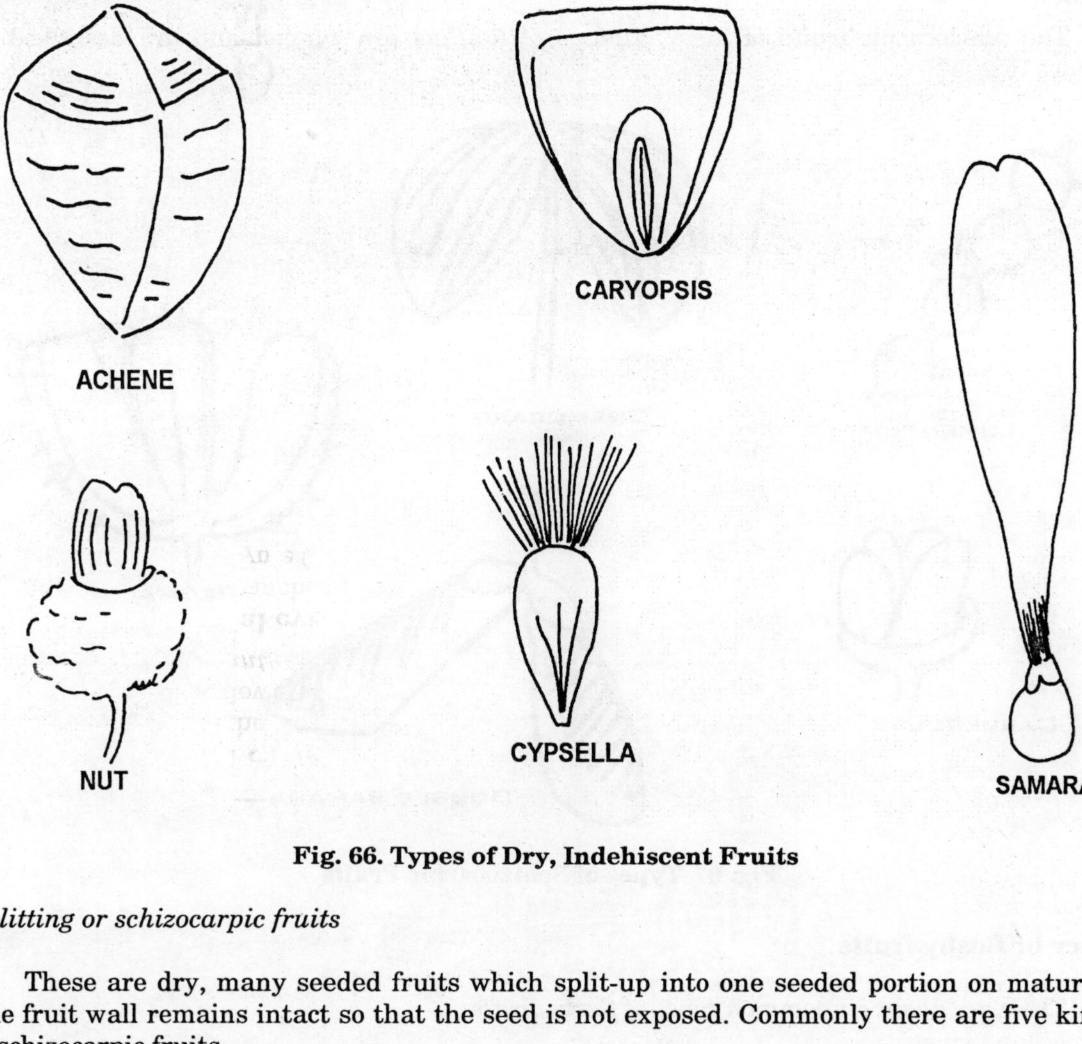

Fig. 66. Types of Dry, Indehiscent Fruits

Splitting or schizocarpic fruits

These are dry, many seeded fruits which split-up into one seeded portion on maturity. The fruit wall remains intact so that the seed is not exposed. Commonly there are five kinds of schizocarpic fruits.

Lomentum is a fruit of leguminous plant which does not dehisce but breaks into one seeded portions at maturity as in Babhul, *Desmodium* etc.

Cremocarp develops from the bicarpellary, syncarpous, inferior ovary with two locules and two ovules. The wall of the fruit has oil ducts. The mature fruit splits from bottom upwards into two, one seeded mericarps that are attached to the stalk - carpophore. This kind of fruit is characteristic of coriander family.

Regma develops from the tricarpellary, syncarpous, superior ovary which is trilocular and has 3 to 6 ovules attached to the central axis. The mature fruit splits into three, one or two seeded portions as in Castor. These portions are called cocci (sing. Coccus).

Carcerule or Carcerulus develops from the bicarpellary, syncarpous, superior ovary which is 4-locular and 4 ovuled. The four locules with one seed each separate at maturity. This kind of fruit is seen in the members of the family Lamiaceae (Labiatae). such as *Ocimum*.

The schizocarpic fruits of *Acer, Hiptage, Ailanthus* are winged and are described as samara (Fig. 67).

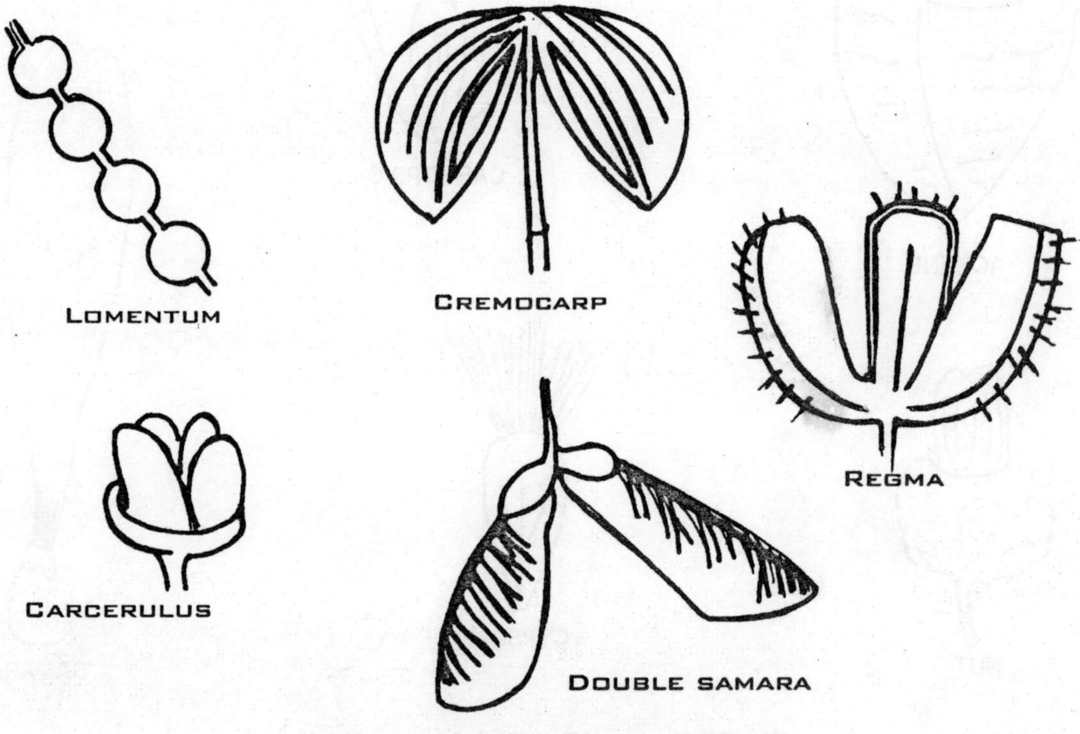

Fig. 67. Types of Schizocarpic Fruits

Juicy or fleshy fruits

There are about six common types of fleshy fruits.

Drupe develeps from polycarpellary, syncarpous, superior or inferior ovary and bears one to few seeds. The fruit is distinctly 3 layered, the outermost epicarp, the middle one mesocarp and the innermost endocarp. Epicarp is usually thin but protective cover, the mesocarp is fleshy, juicy or fibrous and the endocarp is thick and hard protecting the seed inside. Drupes are found in mango, *Ziziphus*, coconut etc.

Bacca or Berry is usually a many seeded fruit developing from a polycarpellary, syncarpous, superior or inferior ovary. Excepting the seeds, all other parts of the fruit are fleshy or juicy. Tomato, guayava, brinjal, grapes are some of the common examples of berries.

Pepo is many seeded, fleshy fruit developing from a tricarpellary, syncarpous, inferior ovary with parietal placentation. This is a characteristic fruit of cucumber family.

Pome develops from two to many carpellary, syncarpous, multilocular, inferior ovary. The fruit is often enclosed by a fleshy receptacle. The edible portion of an apple is really this fleshy receptacle (a false fruit) surrounding the non- edible true fruit.

Hespiridium develops from the polycarpellary syncarpous, superior ovary which is many locular, and also many seeded. Placentation is axile. The thin endocarp approaches the central axis and thus forms many locules. The epicarp and mesocarp are fused. The juicy edible part are the tubular hairs of the endocarp. This kind of fruit is characteristic of lemon group.

Balausta is a special kind of fruit developing from the polycarpellary, syncarpous, inferior ovary which is many loculed with axile placentas in the upper half and one loculed with parietal placentas in the lower half of the ovary. The epicarp is leathery and the testa of the seeds is juicy. Eg. Pomogranate (Fig. 68).

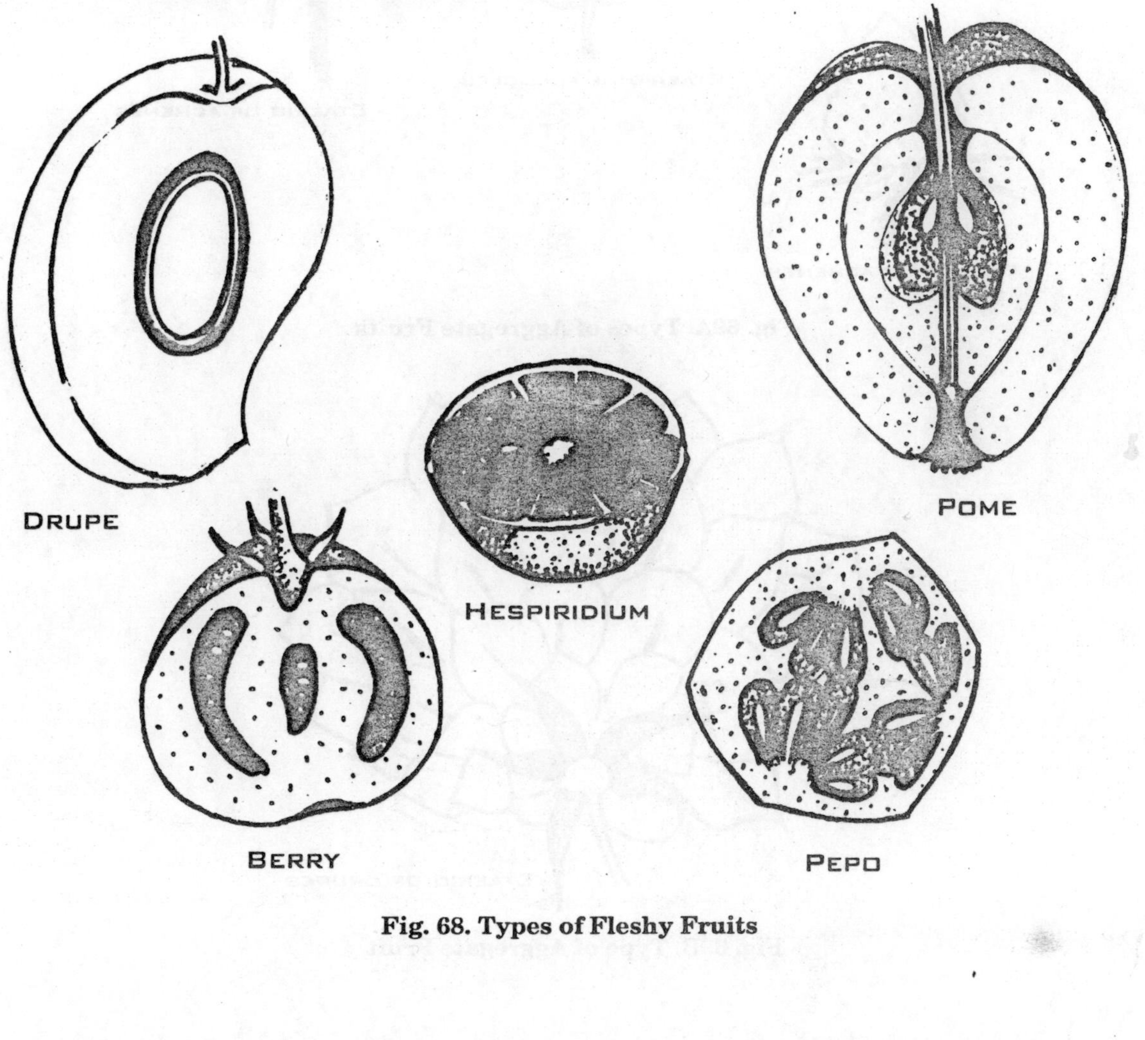

Fig. 68. Types of Fleshy Fruits

2. Aggregate fruits

These fruits develop from polycarpellary apocarpous gynoecium. Each of the distinct carpel forms a fruitlet so that there is a cluster of fruitlets in a single aggregate fruit. The individual fruitlets may be of various kinds such as follicles, achenes, berries and so on. The aggregate fruit is termed accordingly as an etaerio of follicles as in Michelia, etaerio of berries as in custard apple, etaerio of achenes as in *Clematis* etc. (Fig. 69).

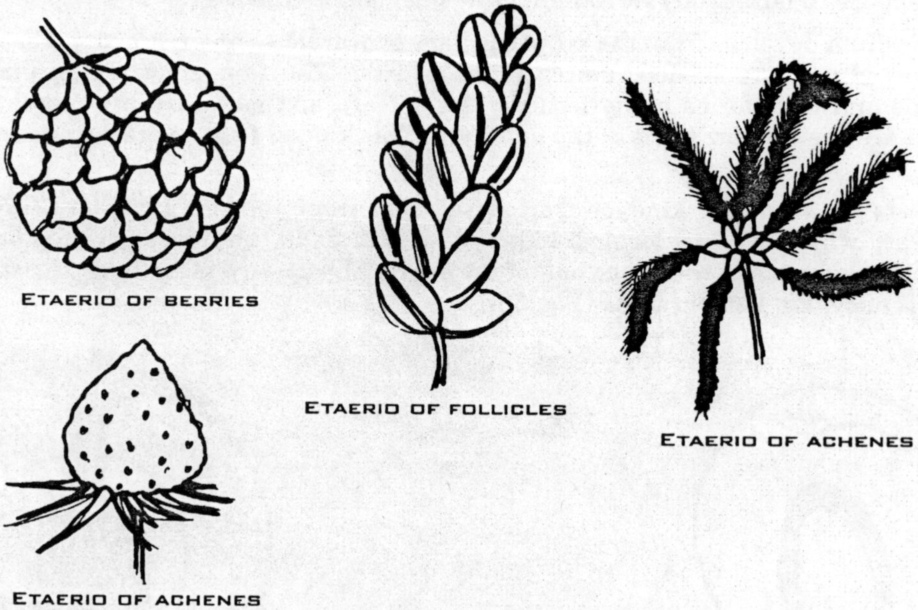

Fig. 69A. Types of Aggregate Fruits

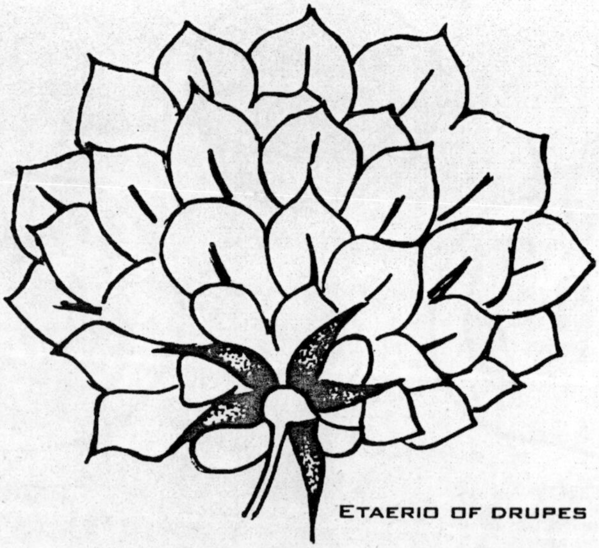

Fig. 69B. Type of Aggregate Fruit

3. Composite or multiple fruits

These fruits develop from the inflorescence. The individual flowers are either very closely placed in the inflorescence as in mulberry or very minute as in fig. The fruit of mulberry, jack fruit, pineapple etc. is termed as sorosis while that of fig is termed as syconus. Thus sorosis develops from the spike inflorescence whereas syconus develops from the hypanthodium inflorescence (Fig. 70).

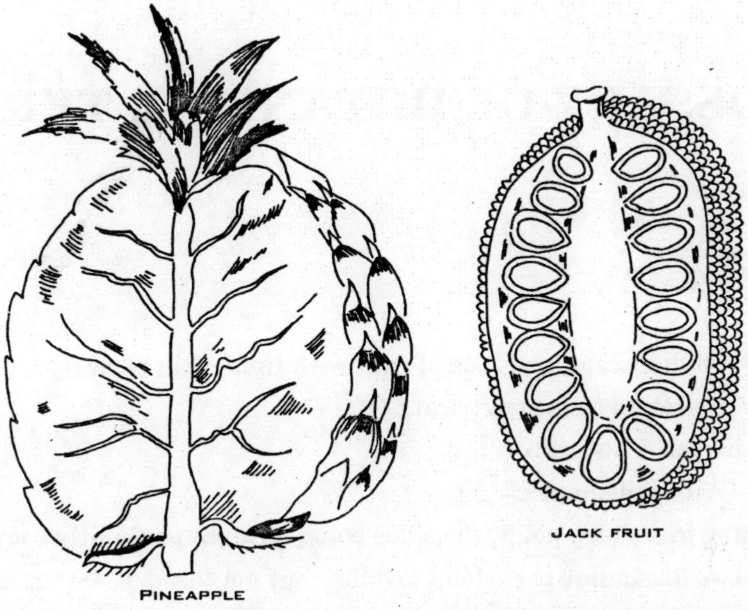

Fig. 70A. Types of composite fruits

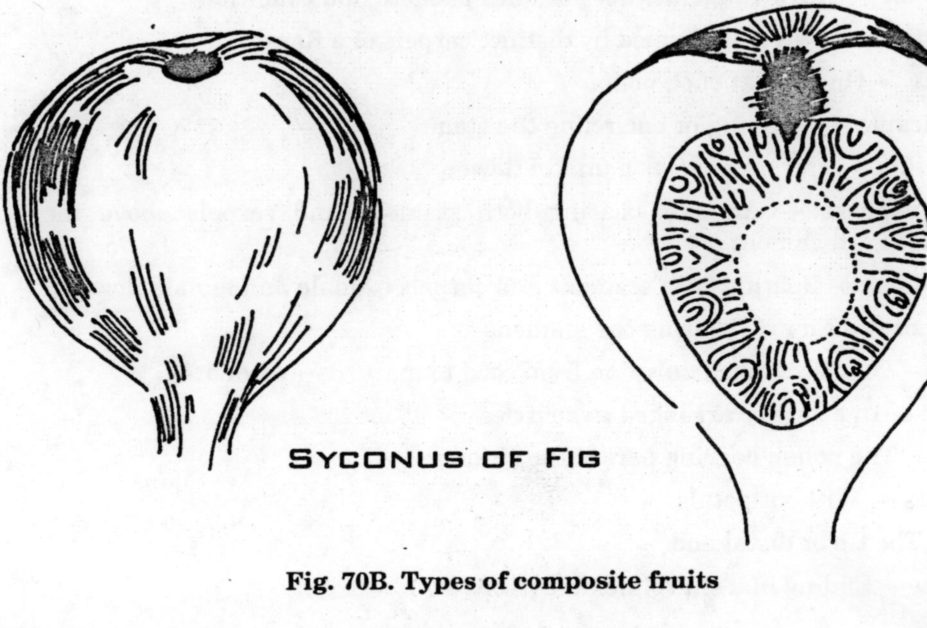

Fig. 70B. Types of composite fruits

2

GLOSSARY OF BOTANICAL TERMS

- Achene — Dry, indehiscent, one seeded fruit with tight, thin pericarp.
- Actinomorphic — Regular, symmetrical
- Acumen — An apex drawn into a tail.
- Acuminate — With tailed apex.
- Acute — Ending in a sharp point, the sides being straight or slightly convex.
- Adherent — Two dissimilar parts held together but not fused.
- Adnate — Two dissimilar parts united or fused.
- Aestivation — The arrangement of perianth parts in bud condition.
- Aggregate fruit — A fruit formed by distinct carpels of a flower.
- Alternate — One leaf at each node.
- Amplexicaule — Clasping or encircling the stem.
- Androecium — The stamens as a unit of flower.
- Androgynophore — An axis bearing both stamens and carpels above the point of perianth attachment.
- Androgynous — Bearing both stamens and carpels or male and female flowers.
- Androphore — An axis bearing the stamens.
- Annual — Of one season's duration from seed to maturity and death.
- Annular — In a ring or arranged in a circle.
- Anther — The pollen bearing part of the stamens.
- Apetalous — Without petals.
- Apex — The tip or distal end.
- Apiculate— Ending in a sharp, flexible point.

- Apocarpous — With separate carpels.
- Areole — The open space formed by anastomosing veins.
- Aril — An appendage or an out growth of the seed coat, growing out from the hilum or funiculus.
- Aromatic — With aroma or scented.
- Articulate — Jointed, provided with nodes.
- Auricle — An ear-shaped appendage or part, usually at the base of leaves or petals.
- Autotrophic — Self sufficient about food.
- Awn — A bristle - like appendage.
- Axil — Upper angle between stem and a petiole.
- Axile — Belonging to axis.
- Axillary — In an axil.
- Basal — Situated at the base.
- Basilar — Attached at the base.
- Berry — Fleshy indehiscent, few or many seeded fruit, like tomato or grape.
- Bicarpellary — Having two carpels.
- Bicarpellate — Having two carpels.
- Bifurcate — Divided into two branches
- Bilocular — With two locules.
- Bisexual — Having both sexes present and functional in a flower.
- Biternate — Twice divided into three parts.
- Bladdery — Inflated like a bladder.
- Blade — The expanded part of the leaf.
- Bract — A much reduced or modified leaf, usually subtending a flower in the axil.
- Bracteate — Bearing bract.
- Bracteolate – Bearing bracteole
- Bracteole — A secondary bract.
- Bulb — An underground organ formed by thickened leaf-bases arranged compactly on a reduced, disc -like stem.
- Bulbil — A small bulb, usually in the leaf axils.
- Calyptrate — Resembling a lid-like structure.
- Calyx — The outer whorl of a flower composed of sepals.
- Campanulate — Bell shaped.
- Capitate — Formed like a head or a compact cluster.
- Capitulum — A dense, head-like inflorescence subtended by a whorl of bracts.

- Capsular — Capsule-like.
- Capsule — A dry, dehiscent, few to many seeded fruit.
- Carcerule — A fruit formed of few to many united carpels which separate at maturity.
- Carcerulus — Carcerule-like fruit.
- Carpel — One of the members of gynoecium, the innermost whorl of a flower. It consists of an ovule-bearing swollen basal ovary, slender leafless style which terminates into a stigma.
- Caryopsis — A fruit of grasses, dry, indehiscent, one seeded fruit in which the epicarp is fused with the seed coat.
- Catkin — An inflorescence with long axis bearing unisexual, sessile flowers.
- Caudate — Bearing a tail-like appendage.
- Chaff — A small, thin, dry, membranous bract or a glume.
- Chaffy — Like chaff.
- Cilia — Hairs borne along margins.
- Ciliate — Bearing cilia.
- Columnar — Column-like, pillar-like.
- Compound leaf — A leaf of two or more leaf-lets or of one leaflet jointed on the petiole.
- Compressed — Flattened from sides.
- Conduplicate — Folded together lengthwise.
- Cone — A dense and elongated collection of flower or fruits with bracts on a central axis.
- Confluent — Merging or blending together.
- Connate — United or joined. Applied to the union of two or more similar parts or organs.
- Contorted — Convolute, twisted.
- Cordate — Heart-shaped, with a sinus and rounded lobes at the base.
- Coriaceous — Of leathery texture.
- Corm — A solid, globose, underground stem.
- Corolla — Second whorl of a flower comprising petals.
- Corona — Any appendage between the corolla and stamens.
- Corymb — Short and broad, more or less flat topped indeterminate inflorescence, the outer flowers opening first.
- Corymbose — Corymb-like.
- Cotyledon — Seed leaf, The primary leaf or leaves in the embryo.
- Cremocarp — A dry, dehiscent, two seeded fruit of the Coriander Family, each half a mericarp borne on a hair-like carpophore.
- Crenate — Obtusely toothed, convex toothed.
- Crustaceous — Dry, membraneous.
- Cupular — Cup-shaped.

- Cyathium — A type of inflorescence characteristic of *Euphorbia,* a female flower represented by single ovary on a stalk surrounded by several male flowers, each represented by a single, stalked stamens, all enclosed within a bracteate envelope.
- Cyme — A broad, more or less flat topped, determinate inflorescence with central flower opening first.
- Cymose umbel — Determinate umbel inflorescence with central flowers maturing first.
- Deciduous — Falling at the end of one season of growth.
- Dehiscence — The method or process of opening of a fruit or anther.
- Dehiscent — One which opens at maturity.
- Dentate — With acute or concave teeth.
- Depressed — Pressed down from the top.
- Diadelphous — Forming two bundles or clusters.
- Dichotomous — Forked, two branched from a point.
- Didynamous — With four stamens, in two pairs of two different lengths, two short two long.
- Diffuse — Loosely branching or spreading.
- Digitate — Palm-like compound with the members arising from one point.
- Dioecious — Having male and female folwers on separate plants.
- Disc — A more or less fleshy development of the receptacle, the receptacle in the head of sunflower family, a flattened end, sometimes spelled disk.
- Dissected — Divided into many narrow parts.
- Distichous — Two-ranked.
- Distinct — Separate.
- Dorsiventral — Flattended surface with definite upper and lower surface.
- Double — Said of flowers that have more than the usual number of perianth members.
- Drupaceous — Drupe-like.
- Drupe — A fleshy, one seeded, indehiscent fruit with seed enclosed in a hard stone or endocarp.
- Ebracteate — Without bract.
- Echinate — With stout prickles.
- Elongate — Lengthened, which has become long.
- Emerginate — With a shallow notch at the apex.
- Entire — With a continuous margin.
- Epicarp — Outer skin or cover.
- Epiphytic — A plant growing on another plant or on some other elevated support.
- Erect — Upright.

- **Exserted — Projected beyond.**
- **Exstipulate — Without stipules.**
- **Fascicle — Bundle.**
- **Filament— Thread, stalk of the stamen.**
- **Filiform — Thread like.**
- **Fimbriate — Fringed, with ciliate margin.**
- **Flexuous — Zigzag.**
- **Floret — Small flower, an individual flower of the sunflower or grass family.**
- **Foliaceous — Leaf-like.**
- **Foliolate — With leaflets.**
- **Follicle — Dry, deshiscent many seeded fruit opening only on one suture.**
- **Free-central — Said of placentation when ovary is one-locular and ovules are attached to the central column.**
- **Frond — Leaf of a fern.**
- **Gamopetalous — With united petals.**
- **Gamosepalous — With united sepals.**
- **Glabrous – Without hairs**
- **Gland-dotted — Having glands**
- **Glaucous — Greyish-green, bluish-green.**
- **Globose — Spherical, globe-shaped.**
- **Glochid— A minute bristle.**
- **Glume — Dry or chaffy membraneous bract.**
- **Gonous — Angular.**
- **Gynobasic — Arising from the base of the gynoecium.**
- **Gynoecium — The female element of a flower, composed of one or many carpels.**
- **Gynophore — Stalk of the ovary.**
- **Herb — Plant without firm woody structure.**
- **Herbaceous — Non-woody.**
- **Hypanthium — A cup-like receptacle, usually derived from the perianth and stamens, on which are seemingly borne calyx, corolla and stamens.**
- **Hypanthodium — A special type of inflorescence in *Ficus* with vessel like receptacle having small opening at the top and enclosing several minute unisexual flowers. It ripens into a fruit termed as syconus.**
- **Hypogynous — Below the ovary.**

- Imbricate — Overlapping, applied to the arrangement of perianth parts in bud condition.
- Indehiscent — Not breaking or opening even at maturity.
- Inferior — Said of the ovary when it is situated below the perianth.
- Inflorescence — Collection of flowers on a special leafless branch.
- Internode — The part of an axis between two successive nodes.
- Involucral — Of involucre.
- Involucre — Whorl of bracts.
- Isobilateral — Having two similar surfaces.
- Keel — The two front petals, united to form a boat-shaped structure of a papilionaceous corolla.
- Lamellar — Provied with many fin-like partitions.
- Lanceolate — Lance-shaped much longer than broad, widening above the base and tapering to the apex.
- Latex — Milky sap.
- Lax — Loose, opposite of congested or compact.
- Leaflet — One part of a compound leaf.
- Legume — Simple, dry fruit, dehiscing on both sutures, derived from monocarpellary gynoecium with marginal placentation.
- Ligulate — Tongue-shaped, strap-shaped.
- Ligule — A strap-shaped organ.
- Linear — Long and narrow.
- Lip — One of the parts in an unequally devided corolla or calyx, these parts are usually two, the upper lip and lower-lip.
- Lobe — A segment of an organ.
- Locule — Compartment.
- Loment — A leguminous fruit contracted between the seeded segments separating at maturity.
- Mamillate — Rough, studded with rounded projections.
- Marginal — Said of placentation where it is situated along one margin of the ovary.
- Monadelphous — Forming one bundle.
- Moniliform — Constricted laterally and appearing bead-like.
- Monocarpellary — With single carpel.
- Monoecious — Male and female flowers borne on the same plant.
- Multicarpellary — With many carpels.
- Multilocular — With many locules.

- Muricate — A surface with small blunt projections.
- Napiform — Of the shape of a beet root.
- Node — A joint of a stem where a leaf is borne.
- Nut — Dry, indehiscent, one seeded fruit.
- Oblanceolate — The reverse of lanceolate, broad towards distal end and tapering towards base.
- Oblong — Long, narrow, with more or less parallel margins.
- Obovate — The reverse of ovate, broader at distal end, narrowed at base.
- Obreniform — The reverse of reniform, inverted, kidney shaped.
- Ocreate or Ochreate — With ocrea or nodal sheath formed by fusion of two stipules.
- Operculum — Lid or cover.
- Opposite — Two at a node, on opposing sides of an axis.
- Orbicular— Circular or disc-shaped.
- Ovary — Ovule-bearing part of a carpel.
- Ovate — Egg-shaped.
- Ovule — The body which after fertilization develops into seed.
- Palea — Bract of the individual floret of Asteraceace and Poaceae.
- Paleaceous — Bearing palea.
- Palmate — Palm — shaped.
- Panicle — Branching, raceme.
- Paniculate — Panicle-like.
- Papilionaceous — Type of the corolla which appears like a butterfly. It has five petals, the largest one is termed standard or vexillum, two lateral wings and two smallest often united to form a keel.
- Pappus — Peculiar modified calyx of the sunflower family.
- Parasite — A plant growing on other plants and depending for its food on them.
- Parietal — A type of placentation where the placentas are borne on the walls of the ovary.
- Pedicel — Stalk of a single flower in an inflorescence.
- Peduncle — Stalk of the inflorescnce or a solitary flower.
- Peltate — Flat topped with centrally attached stalk.
- Pendulous — Hanging from above.
- Pentacarpellary or Pentacarpellate — Composed of five carpels.
- Pentalocular — Five chambered.
- Pentamerous — With parts in multiples of five.

- Pepo — A fleshy, indehiscent fruit of cucumber family.
- Perennial — Of three or more season's duration.
- Perianth — A collective term for calyx and corolla when any one or both of them are present.
- Persistent — Remaining attached, not falling off.
- Petal — Individual member or a lobe of corolla.
- Petaloid — Petal like.
- Petiole — Stalk of leaf,
- Pinna — A primary division or a leaflet of pinnate leaf.
- Pinnae — Plural of pinna.
- Pinnate — Feather shaped, with the leaflets of a compound leaf placed on either side of rachis.
- Pinnatifid — A simple leaf shallowly lobed in a pinnate mannar.
- Pinnatisect — A simple leaf deeply divided to the midrib in a pinnate mannar.
- Placenta — A place or part in the ovary where ovules are attached.
- Placentation — The arrangement of placentas in the ovary.
- Plicate — Folded lengthwise.
- Plumose — Feather-like with fine hairs.
- Pod — A fruit of legume family.
- Pollen — Spores or grains borne by the anther.
- Pollination — Transfer of pollen from the anther to the stigma.
- Pollinia — Pollinium Sing. Sacs containing waxy mass of pollen inside the anther.
- Polypetalous — Corolla or flower with free petals.
- Pore – Small opening
- Prickle — Sharp and stout outgrowth on the stem, petiole etc. as in rose.
- Prickly — With prickles.
- Procumbent — Trailing or lying flat, but not rooting.
- Prostrate — Lying flat on the ground.
- Pseudowhorls — False whorls.
- Pyriform — Pear shaped.
- Raceme — A single, elongated, indeterminate inflorescence with pedicelled flowers.
- Racemose umbel — Umbel with outer flowers opening first.
- Rachis — Axis bearing flowers or leaflets.
- Ray — Outer florets of some Asteraceae with an extended or ligule-like part of the corolla, also a branch of an umbel.

- Receptacle — A more or less enlarged end of the flower axis bearing parts, an enlarged end of an inflorescence axis in Asteraceae, hypanthodium inflorescence.
- Reniform — Kidney-shaped.
- Reticulate — Net-like.
- Rhizome — An underground stem with scale leaves, as in ginger.
- Rugose — Covered with wrinkles.

- Saccate — Sac-like, pouch-like.
- Sagittate – Arrow -head shaped.
- Samara — Dry, indehiscent, one seeded fruit with broad wings.
- Samroid — Winged fruit.
- Saprophytic — Living on dead organic matter.
- Scandent — Climbing but without tendrils.
- Scape — Leafless peduncle arising from the ground.
- Scapigerous — Producing scape.
- Scarious — Thin, dry, membraneous.
- Schizocarpic — Splitting but not dehiscing, seprating into one seeded parts at maturity.
- Secund — One sided.
- Semi-inferior — Half inferior, applied to ovary which is partly above and partly below the calyx, corolla and stamens.
- Sepaloid — Sepal-like.
- Serrate — Applied to leaf margin with the teeth pointing forward.
- Serrulate — Minutely serrate.
- Sessile — Without stalk.
- Sheath — Long and tubular structure.
- Shrub — A woody plant branching from the base.
- Sori — Plural of sorus.
- Sorus — A cluster of fruiting bodies (spores) of ferns.
- Spadix — A thick, fleshy axis bearing sessile flowers.
- Spathaceous — Spathe-like.
- Spathe — A large bract subtending an inflorescence, particularly spadix.
- Spatulate — (or spathulate) — Spoon shaped.
- Spike — An indeterminate inflorescene with long axis and sessile flowers.
- Spinous — Bearing spines.
- Spinulose — Bering minute spines.
- Spore — A simple, grain like reproductive body of ferns and other lower plants.

- Spur — A tubular or sac-like projection.
- Spurred — Bearing spur.
- Stamen — The pollen bearing organ of a seed plant having a narrow filament and an anther.
- Staminode — A sterile stamen, stamen without anther or pollen grains.
- Stem— Main axis of a plant bearing branches, leaves, flowers and fruits.
- Sterile — Lacking functional sex organs.
- Stigma — The part of the carpel that receives pollen.
- Stipulate — Bearing stipules.
- Stipule — A basal appendage of the petiole.
- Style — Slender, seedless part of the carpel that ends in a stigma.
- Sub-capitate — Slightly capitate.
- Sub-opposite — More or less opposite.
- Sub-sessile — More or less sessile.
- Sub-peltate — More or less peltate.
- Subulate — Narrow and tapering from base to apex.
- Succulent — Juicy, fleshy, soft and thick.
- Superior — Said of the ovary when calyx, corolla, and stamens arise from below it.
- Superposed — One above the other.
- Syncarpous — With united carpels.

- Tendril — A thread — like twining structure or organ.
- Terete — More or less cylindrical.
- Terminal — At the tip, apical or distal end.
- Terrestrial — Of the ground.
- Testa — Outer seed coat.
- Tetracarpellary or Tetra carpellate — With four carpels.
- Tetradynamous — An androecium of six stamens of which four are long two short.
- Tetralocular — Four chambered.
- Tetramerous — Floral parts in multiples of four.
- Thyrsoid panicle — A panicle inflorescence with main axis indeterminate, branches determinate.
- Tomentose — Densely soft hairy.
- Torulose —Irregularly swollen at close intervals.
- Torus — Receptacle.
- Tree — A woody plant with a clear trunk below and branches above.

- **Tricarpellary or Tricarpellate — With three carpels.**
- **Trilocular — Three chambered.**
- **Trimerous — Floral parts in multiples of three.**
- **Truncate — Nearly or quite straight across.**
- **Tuberculate — Covered with rounded protruding bodies.**
- **Turgid — Swollen from fullness.**

- **Umbel — An indeterminate inflorescence in which flowers appear to be arising from a single point and forming an umbrella-like shape.**
- **Umbellet or Umbellule — Small umbel.**
- **Unilocular — One chambered.**
- **Unisexual — Bearing one sex, either male or famale.**

- **Valvate — Said of aestivation where the parts meet by their edges, not over-lapping.**
- **Verticil — A whorl.**
- **Verticillate — Arranged in whorls.**
- **Verticillaster inflorescence — One with the flowers in whrols, the whorls remote from one another, the flowers themselves sessile. These whorls are false whorls. They are actually sessile cymes in the axils of opposite leaves or bracts.**

- **Whorled — Arranged in a whorl or circle.**
- **Woody — Provided with wood.**

- **Zygomorphic — Symmetrical in one plane only, usually along, anterio-posterior line, petals are unequal in size and shape.**

3

KEY TO IDENTIFICATION OF FAMILIES OF MEDICINAL PLANTS

KEY TO THE FAMILIES

1a. Plants non-green, saprophytic; plant body umberlla-like .... **Agaricaceae** (Chhatra Kul)

1b. Plants green, autotrophic, if rarely non- green then either flower - bearing or epiphytic on rocks, tree bark etc.... 2

2a. Plants non - flowering, without seeds.... 3

2b. Plants flowering, or cone - bearing with seeds. 4

3a. Plants terrestrial, having roots, stem and leaves; leaves half - moon shaped, spore - bearing. .... **Adiantaceae** (Hansapadi Kul)

3b. Plants often epiphytic on wet rocks, not having roots, stem or leaves........ **Parmeliaceae** (Dagadful Kul)

4a. Plants cone- bearing, seeds not enclosed in fruits........ 5

4b. Plants with typical flowers, seeds enclosed in fruit.... 8

5a. Leaves long, narrow.... 6

5b. Leaves short, scale like... 7

6a. Leaves needle like, 2 to many together, seeds many, dry .... **Pinaceae** (Devdar Kul)

6b. Leaves flat, often, with two whitish bands beneath, solitary, seeds one or two, fleshy or juicy. .... **Taxaceae** (Talispatra Kul)

7a. Branches alternate, leaves densely crowded, seeds in globose cones .... **Cupressaceae** (Hauber Kul)

7b. Branches opposite or whorled, leaves laxly arranged, seeds fleshy in elongate cones. .... **Ephedraceae** (Somlata Kul.)

8a. Leaves often dorsiventral and raticulately veined, perianth when present, usually arranged in whorls of 5 or 4 if in whorls of 3, then plants woody, cotyledons mostly two 9

8b. Leaves often isobilateral with parallel venation, perianth when present, usually in whorls of 3, less commonly less than or more than 3 but then plants herbaceous, cotyledons mostly one or undifferentiated..........152

9a. Perianth present 10

9b. Perianth absent..151

10a. Perianth in two or more whorls and clearly differentiated into calyx and corolla..........11

10b. Perianth in one whorl, if in two or more whorls then not differentiated into calyx and corolla..125

11a. Corolla of seprate petals..12

11b. Corolla of united petals 86

12a. Stamens numerous, more than 15 13

12b. Stamens few, less than 15 40

13a. Gynoecium of usually many, free carpels 14

13b. Gynoecium of one or two to many united carpels 17

14a. Plants growing in water .... **Nelumbonaceae** (Pundarik Kul.)

14b. Plants growing on land or else where..15

15a. Plants herbaceous.................... **Ranunculaceae** (Bachnag Kul.)

15b. Plants woody ....---16

16a. Flowers showy, stamens inserted on hypanthium or calyx cup............ **Rosaceae** (Gulab Kul.)

16b. Flowers not showy, stamens inserted on the receptacle.............. **Annonaceae** (Sitaphal Kul.)

17a. Stamens monadelphous or in 4-6 bundles..18

17b. Stamens distinct or nearly so..24

18a. Ovary monocarpellery, unilocular, fruit a legume or loment........... **Mimosaceae** (Babul Kul.)

18b. Ovary 2 to many carpellary, 2 to many locular, fruit not a legume or loment..........19

19a. Stamens monadelphous..20

19b. Stamens in 4-6 bundles 22

20a. Anthers 1 - celled, staminodes absent 21

20b. Anthers 2- celled, staminodes often present .... **Sterculiaceae** (Muchkund Kul)

21a. Leaves digitate.................... **Bombacaceae** (Shalmali Kul.)

21b. Leaves simple.................... **Malvaceae** (Karpas Kul.)

22a. Flowers 5- merous, fruit dry..23

22b. Flowers 2, 3, 4 or 6 merous, fruit fleshy .... **Clusiaceae** (Kokam Kul)

23a. Leaves digitate, fruit dehiscent.................... **Bombacaceae** (Shalmali Kul.)

23b. Leaves simple, fruit indehiscent .... **Dipterocarpaceae** (Sarg or Ral Vruksha Kul.)

24a. Trees with yellow latex.................... **Clusiaceae** (Kokam Kul.)

24b. Trees, shrubs or herbs, with or without yellow latex 25

25a. Sepals 2 or 3 26
25b. Sepals 4,5 or more........ 28
26a. Leaves fleshy, placentas free-central or basal 27
26b. Leaves not fleshy, placentas parietal .... **Papaveraceae** (Ahiphen Kul.)
27a. Flowers terminal, often clustered, seeds few .... **Portulacaceae** (Lonica or Ghol Kul.)
27b. Flowers axillary, often clustered, seeds few........ **Aizoaceae** (Vasu or Vasuk Kul.)
28a. Ovary superior 29
28b. Ovary inferior or sunk in thalamus 37
29a. Placentas parietal or marginal 30
29b. Placentas axile or lamellar........ 33
30a. Leaves pinnately compound, placentas marginal,
fruit a legume or loment........ **Mimosaceae** (Babul Kul.)
30b. Leaves simple or digitate, placentas parietal, fruit not a legume or loment 31
31a. Petals 5, ovary sessile or subsessile, fruit capsular........ 32
31b. Petals 4, ovary raised on a gynophore, fruits
indehiscent .... **Capparaceae** (Varun or Vayavarna Kul.)
32a. Leaves entire, petals white or rose coloured,
capsules echinate, 2- valved .... **Bixaceae** (Sinduri Kul.)
32b. Leaves palmately lobed, petals golden yellow,
capsules smooth, 5- valved .... **Cochlospermaceae** (Peet karpas Kul.)
33a. Plants growing in water, placentas lamellar 34
33b. Plants growing on land, placentas axile 35
34a. Plants spinous........ **Euryalaceae** (Makhan Kul.)
34b. Plants not spinous .... **Nymphaeaceae** (Kamal Kul.)
35a. Leaves opposite, stamens inserted on the hypanthium .... **Lythraceae** (Mendica Kul.)
35b. Leaves alternate, stamens inserted on the receptacle........ 36
36a. Leaves with sheathing base, fruits fleshy, many seeded........ **Dilleniaceae** (Karambal Kul.)
36b. Leaves not with sheathing base, fruit dry or if fleshy
then few seeded........ **Tiliaceae** (Parushak Kul.)
37a. Leaves gland-dotted........ **Myrtaceae** (Lavang Kul.)
37b. Leaves not gland-dotted 38
38a. Locules of the ovary superposed, seeds with juicy testa........ **Punicaceae** (Dalimba Kul.)
38b. Locules of the ovary not superposed, seeds with dry testa........ 39
39a. Leaves 3-nerved from the base, flowers in axillary fascicles,
petals 5-10, linear- oblong, ovary unilocular, seed one........ **Alangiaceae** (Ankol Kul.)
39b. Leaves one- nerved, flowers in racemes or spikes, petals 4-5, ovoid or
roundish, ovary 2-5 locular, seeds many .... **Lecythidaceae** (Samudraphal Kul.)
40a. Flowers 2-3 merous........ 41
40b. Flowers 4-5 merous........ 45

41a. Plants often twining, flowers unisexual, carpels distinct .... **Menispermaceae** (Guduchi Kul.)
41b. Plants crect, flowers bisexual, carpels united .... **Berberidaceae** (Daruharidra kul.)
42a. Carpel one, fruit a legume or loment 43
42b. Carpels 2-many, fruit not a legume or loment 45
43a. Flowers actinomorphic .... **Mimosaceae** (Babul Kul.)
43b. Flowers zygomorphic 44
44a. Corolla papilionaceous .... **Fabaceae** (Aparajita Kul.)
44b. Corolla not papilionaceous .... **Caesalpiniaceae** (Putikaranj Kul.)
45a. Flowers bisexual 46
45b. Flowers unisexual or polygamous 78
46a. Flowers actinomorphic 47
46b. Flowers zygomorphic 72
47a. Stamens inserted on the receptacle 48
47b. Stamens inserted on the hypanthium or on the top of the ovary 67
48a. Stamens united into a tube 49
48b. Stamens distinct or only basally connate 50
49a. Leaves simple .... **Sterculiaceae** (Muchkund Kul.)
49b. Leaves pinnately compound .... **Meliaceae** (Kadunimb Kul.)
50a. Placentas parietal 51
50b. Placentas axile or ovules pendulous 54
51a. Trees or shrubs 52
51b. Herbs 53
52a. Flowers tetramerous, ovary often raised on a gynophore .... **Capparaceae** (Varun or Vayavarna Kul.)
52b. Flowers pentamerous, ovary not raised on a gynophore .... **Samydaceae**
53a. Stamens 6, tetradynamous, rearly less than 6 .... **Brassicaceae** (Rajika or Rai Kul.)
53b. Stamens more than 6, if 6 then not tetradynamous .... **Cleomaceae** (Tilparni or Tilavan Kul.)
54a. Sepals imbricate in bud 55
54b. Sepals valvate in bud or if imbricate, then trees with pinnately compound leaves 66
55a. Herbs or undershrubs 56
55b. Shrubs or trees 59
56a. Leaves opposite .... **Zygophyllaceae** (Laghu Gokshur Kul.)
56b. Leaves alternate 57
57a. Stamens 4 or 5, fruits drupaceous .... **Vitaceae** (Draksha Kul.)
57b. Stamens 10, fruit capsular 58
58a. Leaves compound, seeds many .... **Oxalidaceae** (Changeri Kul)
58b. Leaves simple, seeds few .... **Linaceae** (Atasi Kul)

59a. Leaves gland dotted .... **Rutaceae** (Sadab Kul)
59b. Leaves not gland- dotted 60
60a. Leaves 2-foliolate .... **Balanitaceae** (Hingan Kul)
60b. Leaves not 2-foliolate 61
61a. Stamens usually 10, if less than 10, then ovary unilocular 62
61b. Stamens 3-5, ovary many locular 64
62a. Ovary 5- locular, ovules many .... **Averrhoaceae** (Karmal or Karmarang Kul.)
62b. Ovary 1-3 locular, rarely 2-5 locular, ovules 1-5 63
63a. Ovary unilocular or rarely 2-5 locular, ovules pendulous .... **Anacardiaceae** (Amra Kul.)
63b. Ovary 3- locular, ovules erect .... **Sapindaceae** (Arishtak Kul.)
64a. Stamens 3, seeds winged .... **Hippocrateaceae** (Saptachakra or Saptarangi Kul.)
64b. Stamens 4-5, seeds not winged 65
65a. Stamens alternating with the petals, fruits capsular .... **Celastraceae** (Jyotishmati Kul.)
65b. Stamens opposite to the petals, fruits drupaceous .... **Vitaceae** (Draksha Kul.)
66a. Leaves simple .... **Rhamnaceae** (Badar Kul)
66b. Leaves, pinnately compound .... **Burseraceae** (Guggulu Kul.)
67a. Ovary superior, although sunk into the calyx-cup, but free from it .... **Lythraceae** (Mendica Kul.)
67b. Ovary inferior or half inferior 68
68a. Plants growing in water, leaves floating, flowers white, fruits 2-horned .... **Trapaceae** (Shrungatak Kul.)
68b. Plants growing on land, leaves and fruits not as above 69
69a. Plants woody, shrubs or trees 70
69b. Plants herbaceous 71
70a. Leaves opposite or sub-opposite, rearly alternate, flowers in spikes or corymbs, fruits one seeded .... **Combretaceae** (Haritaki Kul)
70b. Leaves alternate, flowers in axillary clusters, fruits many-seeded .... **Alangiaceae** (Ankot Kul.)
71a. Rhizome stout, flowers on stout scape, stamens 10, placentas parietal, fruits capsular, seeds many .... **Saxifragaceae** (Pashanbhed Kul.)
71b. Rhizome absent or not stout, flowers in simple or compound umbels, stamens 5, placentas axile, fruits schizocarpic, seeds 2 .... **Apiaceae** (Garger Kul.)
72a. Trees or shrubs 73
72b. Herbs 76
73a. Stamens monadelphous, ovary 5-locular, fruit dry formed of twisted follicles .... **Sterculiaceae** (Muchkund Kul.)
73b. Stamens distinct, ovary 2-3 locular, fruits not as above 74
74a. Trees or shrubs, placentas parietal, fruits dehiscent 75
74b. Herbaceous climbers, placentas axile, fruit bladdery .... **Sapindaceae** (Arishatak Kul.)

75a. Leaves pinnately compound, stamens 5-10,
ovary 3- carpellary, seeds winged .... **Moringaceae** (Shigru Kul.)

75b. Leaves simple, stamens 4-6, ovary 2-carpellary,
seeds not winged .... **Capparaceae** (Varun Kul.)

76a. Flowers not in umbels, ovary superior 77

76b. Flowers in simple or compound umbles, ovary inferior .... **Apiaceae** (Garger Kul)

77a. Leaves much dissected, sepals 2 .... **Fumariaceae** (Parpat Kul.)

77b. Leaves simple, entire or serrate, sepals 4-5 .... **Violaceae** (Kampushpa Kul.)

78a. Ovary inferior 79

78b. Ovary superior 80

79a. Erect herbs, flowers in umbels, ovary 2-locular, fruit a
cremocarp .... **Apiaceae** (Garjar Kul)

79b. Climbing herbs, flowers in axillary clusters or racemes,
ovary unilocular, fruit pepo or capsular .... **Cucurbitaceae** (Karkati Kul.)

80a. Carpels loosely united, maturing into 2-6 distinct
samaras .... **Simaroubaceae** (Maharukh Kul.)

80b. Carpels united, fruits not as above 81

81a. Placentas parietal 82

81b. Placentas axile, ... 83

82a. Leaves lanceolate, entire, petals with a scale at
base, fruits globse, hairy .... **Flacourtiaceae** (Prachinamalak Kul.)

82b. Leaves palmately lobed, serrate, petals without
scale, fruits ovoid, glabrous .... **Caricaceae** (Erand Karkati Kul.)

83a. Ovary 5-locular .... **Sterculiaceae** (Muchkund Kul.)

83b. Ovary 1-3 locular, rearly more than 3-locular but then fruits one seeded 84

84a. Leaves pinnately veined, flowers polygamous 85

84b. Leaves pinnately or palmately veined, flowers strictly
unisexual .... **Euphorbiaceae** (Erand Kul.)

85a. Ovary 3-locular, fruit 3-seeded .... **Sapindaceae** (Arishtak Kul.)

85b. Ovary 1 or 2-5 locular, fruits 1 seeded .... **Anacardiaceae** (Amravruksha Kul.)

86a. Flowers unisexual or polygamous 87

86b. Flowers bisexual 90

87a. Ovary superior, 2 or more locular with axile placentas 88

87b. Ovary inferior, unilocular with parietal or basal placentas 89

88a. Leaves exstipulate, calyx enlarging in fruit,
ovary 4-8 locular .... **Ebenaceae** (Tinduk Kul.)

88b. Leaves stipulate or exstipulate, calyx not enlarging in fruit,
ovary 2-3 locular .... **Euphorbiaceae** (Erand Kul.)

89a. Tendril climbers, flowers unisexual, in axillary fascicles or
racemes, placentas parietal .... **Cucurbitaceae** (Karkati Kul)

89b. Not tendril climber, flowers polygamous in capitulum inflorescence, placenta basal .... **Asteraceae** (Shatpatri Kul.)
90a. Fruit a legume or loment .... **Mimosaceae** (Babul Kul.)
90b. Fruit not a legume or loment 91
91a. Ovary inferior 92
91b. Ovary superior or semi-inferior 100
92a. Stem parasites, calyx ring-like .... **Loranthaceae** (Bandak Kul.)
92b. Plants autotrophic, calyx various 95
93a. Flowers in capitulum inflorescence, ovary unilocular .... **Asteraceae** (Shatapatri Kul.)
93b. Flowers not in capitulum inflorescence, ovary 1-10 locular 94
94a. Leaves opposite or whorled 95
94b. Leaves alternate 97
95a. Leaves stipulate, ovary 2-4 locular, rearly unilocular, ovules few to many 96
95b. Leaves exstipulate, ovary 3-locular, locules one-ovuled .... **Valerianaceae** (Jatamansi Kul.)
96a. Tree, flowers in dense globose heads, petals imbricate, ovaries confluent, ovules many in each locule .... **Naucleaceae** (Kadamb Kul.)
96b. Habit various, flowers, petals and ovaries not as above, if ovaries confluent, then ovule solitary in each locule .... **Rubiaceae** (Manjishtha Kul.)
97a. Plants woody, flowers actinomorphic, anthers distinct 98
97b. Plants herbaceous or soft- wooded, flowers zygomorphic, anthers united .... **Lobeliaceae** (Nal Kul.)
98a. Leaves gland-dotted, flowers in axillary cymes, corolla calyptrate .... **Myrtaceae** (Eucalyptus, Lavang Kul.)
98b. Leaves not gland- dotted, flowers in spikes or racemes, corolla not calyptrate 99
99a. Spikes or racemes erect, petals white .... **Symplocaceae** (Lodhra Kul.)
99b. Spikes or racemes drooping, petals red .... **Lecythidaceae** (Barringtonia, Samudraphal Kul.)
100a. Ovaries more than one, nearly or quite separate at maturity 101
100b. Ovary one of two to many united carpels 104
101a. Plant juice not milky, flowers often tetramerous, ovaries 4 or 5 .... **Crassulaceae** (Parnabij Kul.)
101b. Plant, juice often milky, flower pentamerous, ovaries 2 102
102a. Pollinia absent .... **Apocynaceae** (Kutaj Kul.)
102b. Pollinia present 103
103a. Filaments distinct, anthers not united with the stigma .... **Periplocaceae** (Sariwa Kul.)
103b. Filaments united into a tube, anthers united with the stigma .... **Asclepiadaceae** (Ark Kul.)
104a. Stamens as many as corolla lobes and opposite to them or more than corolla lobes 105

104b. Stamens as many as corolla lobes and alternating with them or less than corolla lobes 107

105a. Trees, stamens more than corolla lobes, fruits fleshy .... **Sapotaceae** (Madhuk Kul.)

105b. Shrubs, stamens as many as corolla lobed, fruits dry 106

106a. Calyx glandular outside, style branches 5, ovule one on basal placenta .... **Plumbaginaceae** (Chitrak Kul)

106b. Calyx not glandular outside, style one, ovules many on free central placenta .... **Myrsinaceae** (Vidang Kul.)

107a. Ovary unilocular 108

107b. Ovary 2 to many locular 109

108a. Trees, leaves one nerved, fruits fleshy, one-seeded .... **Salvadoraceae** (Pilu Kul.)

108b. Herbs, leaves 3 nerved from the base, fruits capsular, many seeded .... **Gentianaceae** (Bhunimb Kul.)

109a. Flowers actinomorphic or only slightly zygomorphic 110

109b. Flowers zygomorphic 120

110a. Leafless stem parasites, appearing like greenish-yellow threads .... **Cuscutaceae** (Amarvel Kul.)

110b. Plants not as above 111

111a. Stamens 2 .... **Oleaceae** (Parijatak Kul.)

111b. Stamens 4 or more 112

112a. Trees 113

112b. Herbs, undershrubs or shrubs, erect, twining or trailing 115

113a. Leaves alternate or sub- opposite .... **Ehretiaceae** (Bahuvar Kul.)

113b. Leaves opposite or whorled 114

114a. Leaves smooth, shining green, 3 nerved from the base .... **Strychnaceae** (Wish-tinduk Kul.)

114b. Leaves rough, dull-green, one nerved .... **Verbenaceae** (Nirgundi Kul.)

115a. Scapigerous herbs with basal leaves, fruits one or two seeded .... **Plantaginaceae** (Ishatgol Kul.)

115b. Plants not as above, fruits two to many seeded 116

116a. Leaves opposite, seeds many 117

116b. Leaves alternate, if opposite then seeds 2-4 118

117a. Plants often shrubby, with milky latex, leaves 1-nerved .... **Apocynaceae** (Kutaj Kul.)

117b. Plants herbaceous, without milky latex, leaves 3-nerved .... **Gentianaceae** (Bhunimb or Chirait Kul)

118a. Fruit a carcerule, splitting into one-seeded portions .... **Boraginaceae** (Bhurundi Kul)

118b. Fruits not as above 119

119a. Plants usually twining or trailing, if erect then fruits 4-6 seeded .... **Convolvulaceae** (Nishottar Kul.)

119b. Plants erect, fruit many seeded .... **Solanaceae** (Kantakari Kul.)

120a. Fruits capsular, seeds many, rarely less then 4 121
120b. Fruits indehiscent or schizocarpic, seeds 2-4 124
121a. Trees with compound, opposite leaves, fruits elongate capsules, seeds winged, rarely leaves simple and fruits fleshy .... **Bignoniaceae** (Arlu Kul.)
121b. Herbs or shrubs with simple, alternate or opposite leaves, capsules small, seeds not winged 122
122a. Leaves often alternate, stipulate, or petioles and pedicles with large sessile glands at their bases .... **Pedaliaceae** (Til Kul.)
122b. Leaves alternate, opposite or rearly whorled, exstipulate, petioles and pedicels without sessile glands at their bases 123
123a. Leaves all opposite or whorled, bracts and often bracteoles coinspicuous, capsules elongate, seeds 2 to many .... **Acanthaceae** (Vasa Kul.)
123b. Leaves upper sometimes alternate, bracteoles often absent or minute, capsules globose or ovoid, seeds numerous. .... **Scrophulariaceae** (Brahmi Kul.)
124a. Ovary entire or 4-lobed, style terminal, stigmas 2, fruit fleshy, indehiscent or carcerulous .... **Verbenaceae** (Nirgundi Kul.)
124b. Ovary deeply 4-lobed, style gynobasic, stigmas 2, fruit carcerulous .... **Lamiaceae** (Tulsi Kul.)
125a. Perianth parts many, distinct, spirally arranged, usually petaloid 126
125b. Perianth parts few, arranged in one or two whorls, sepaloid or petaloid 127
126a. Trees with woody stems and branches, leaves well-developed, gynoecium apocarpous .... **Magnoliaceae** (Champak Kul.)
126b. Shrubs with succulent stems and branches, leaves reduced to spines or absent, gynoecium syncarpous .... **Cactaceae** (Canthari Kul.)
127a. Flowers bisexual 128
127b. Flowers unisexual or polygamous 137
128a. Ovary superior 129
128b. Ovary inferior 136
129a. Leaves with ochreate stipules or plants clinbing by mens of inflorescence bearing tendrils .... **Polygonaceae** (Chukra Kul.)
129b. Leaves exstipulate or stipules and plants not as above 130
130a. Carpels distinct .... **Ranunculaceae** (Clematis, Vatsanabh Kul.)
130b. Carpels united 131
131a. Leafless, wire- like, twining stem parasites .... **Cassythaceae** (Aakashvalli Kul.)
131b. Leaves well-developed, plants autotrophic 132
132a. Perianth lobes united 133
132b. Perianth lobes distinct 134
133a. Herbs or scandent shrubs with showy bracts .... **Nyctaginaceae** (Punarnava Kul.)
133b. Eract shrubs or trees .... **Thymelaeaceae** (Aguru Kul.)
134a. Trecs, leaves often 3-5 nerved from the base, aromatic, flowers in cymes or panicles .... **Lauraceae** (Karpur Kul.)

134b. Herbs or undershrubs, leaves often 1-nerved, not aromatic, flowers in spikes 135
135a. Perianth green .... **Chenopodiaceae** (Palakya Kul.)
135b. Perianth scarious .... **Amaranthaceae** (Apamarg Kul.)
136a. Trees, leaves opposite, flowers actinomorphic with 4 lobed perianth, fruit fleshy .... **Santalaceae** (Chandan Kul.)
136b. Twining herbs or undershrubs, leaves alternate, flowers zygomorphic with unlobed perianth, fruits capsular .... **Aristolochiaceae** (Ishwari Kul.)
137a. Plants growing in water, leaves linear, dichotomously divided .... **Ceratophyllaceae** (Shaival Kul.)
137b. Plants growing on land, leaves not as above 137
138a. Male flowers in cat-kin-like inflorescence 139
138b. Male flowers not in cat-kin-like inflorescence...142
139a. Ovary superior 140
139b. Ovary inferior...141
140a. Fruits fleshy .... **Moraceae** (Wat Kul.)
140b. Fruits dry .... **Corylaceae** (Bhurjpatra Kul.)
141a. Leaves pinnate, exstipulate, fruit drupaceous .... **Juglandaceae** (Akshot Kul.)
141b. Leaves simple, stipulate, fruit a nut, partly or wholly enclosed in a cup .... **Fagaceae** (Mayaphal Kul.)
142a. Ovary 2- many locular 143
142b. Ovary unilocular 145
143a. Ovary superior 144
143b. Ovary semi-inferior.. **Altingiaceae** (Shilaras Kul.)
144a. Ovules 3-6... **Euphorbiaceae** (Erand Kul.)
144b. Ovules 4-20.. **Flacourtiaceae** (Prachinamalak Kul.)
145a. Inflorescene hypanthodium.. **Moraceae** (Wat Kul.)
145b. Inflorescence not hypanthodium 146
146a. Trees or large shrubs 147
146b. Herbs or undershrubs...149
147a. Leaves pinnate with 3-11 leaflets, male flowers with 1-2 sepals .... **Pistaciaceae** (Pista Kul)
147b. Leaves simple, male flowers with 3-5 sepals...148
148a. Leaves 3-nerved from the base, calyx 3-lobed, stamens monadelphous, fruit fleshy but dehiscing into 2-4 valves, seeds with fleshy arils .... **Myristicaceae** (Jatithal Kul.)
148b. Leaves 1-nerved, calyx 3-9 lobed, stamens distinct, fruit dry but indehiscent or samaroid... **Ulmaceae** (Chirbilva Kul.)
149a. Plants dioecious, leaves digitate with 3-11 leaflets .... **Cannabaceae** (Ganja Kul.)
149b. Plants monoecious, leaves simple 150
150a. Perianth green .... **Chenopodiaceae** (Palakya Kul.)

150b. Perianth scarious .... **Amaranthaceae** (Apamarg Kul)

151a. Inflorescence cyathium, ovary 2-3 locular with 2-6 ovules .... **Euphorbiaceae** (Erand Kul.)

151b. Inflorescence not cyathium, ovary unilocular with one ovule 152

152a. Trees with 1-nerved leaves .... **Myricaceae** (Katuphal Kul.)

152b. Herbs or shrubs often with 3-5 nerved leaves .... **Piperaceae** (Marich Kul.)

153a. Perianth well- developed, usually in two series of three each 154

153b. Perianth absent or reduced to scales or hairs 167

154a. Trees or shrubs with fan-like or feather like large stiff leaves .... **Arecaceae** (Narikel Kul.)

154b. Herbs or shrubs, leaves not as above 155

155a. Ovary inferior 156

155b. Ovary superior 164

156a. Flowers zygomorphic 157

156b. Flowers actinomorphic 160

157a. Fertile stamen one or two 158

157b. Fertile stamens 3-6 159

158a. Stamens and styles fused to form a column, seeds minute, numerous .... **Orchidaceae** (Malakand Kul.)

158b. Stamens and styles not fused, seeds rather large, few to many .... **Zingiberaceae** (Haridra Kul.)

159a. Petals dissimilar or only one, fertile stamens 5 or 6 .... **Musaceae** (Kadali Kul.)

159b. Petals more or less similar, fertile stamens 3 .... **Iridaceae** (Gladiolus, Keshar Kul.)

160a. Plants twining, flowers unisexual, fruits winged .... **Dioscoreaceae** (Karand Kul.)

160b. Plants not twining, flowers bisexual, fruit not winged 161

161a. Flowers in spikes, racemes or solitary 162

161b. Flowers in umbels or if solitary then stamens 6 163

162a. Flowers in spike, subtended by brightly coloured bracts, fertile stamens 6 .... **Bromeliaceae** (Ananas Kul.)

162b. Flowers in racemes or solitary, bracts not brightly coloured, fertile stamens 3 .... **Iridaceae** (Keshar Kul.)

163a. Leaves plicate .... **Hypoxidaceae** (Talmuli Kul.)

163b. Leaves not plicate .... **Amaryllidaceae** (Nagdamni Kul.)

164a. Flowers in spathaceous umbels .... **Alliaceae** (Palandu Kul.)

164b. Flowers not in spathaceous umbels 165

165a. Tendril climbers, flowers often unisexual .... **Smilacaceae** (Hiranyashak Kul.)

165b. Not tendril climbers, or if so then flowers bisexual 166

166a. Fertile stamens 6 .... **Liliaceae** (Vanpalandu Kul.)

166b. Fertile stamens 3 .... **Iridaceae** (Keshar Kul.)

167a. Flowers not subtended by dry, chaffy bracts (glumes) 168

167b. Flowers subtended by dry, chaffy bracts (glumes) 169

168a. Leaves often reticulately veined, flowers unisexual or bisexual, arranged on thick, fleshy axis subtended by large spathaceous bract .... **Araceae** (Shuran Kul.)

168b. Leaves parallel-veined with spinulose margins, flowers unisexual, arranged in spikes subtended by leafy bracts .... **Pandanaceae** (Ketki Kul.)

169a Stems usually triangular, leaves with closed sheath, fruit a nut .... **Cyperaceae** (Mustak Kul.)

169b. Stems usually cylindric or flattened, leaves with open sheaths, fruit a caryopsis 170

170a. Stems very tall, woody, covered with large sheaths when young .... **Bambusaceae** (Vaunsha Kul.)

170b. Stems not as above .... **Poaceae** (Trundhanya Kul.)

4

KEY TO IDENTIFICATION OF SPECIES OF MEDICINAL PLANTS

KEY TO THE SPECIES

1. FAMILY - Acanthaceae (VASA Kul.)

1a. Plants spreading on the ground or with weak stem 2
1b. Plants erect or diffuse 5
2a. Corolla with lower lip only 3
2b. Corolla 2- lipped 4
3a. Leaves 2cm. or more long, acute at apex .... *Blepharis maderaspatanus* (Mothi Hadsan)
3b. Leaves less than 2cm. long, obtuse or rounded at apex .... *Blepharis repens* (Hadsan)
4a. Bracts and bracteoles with broad scarious margins - .... *Rungia pectinata*
4b. Bracts and bracteoles narrow, without scarious margins .... *Justicia procumbens* (Ghati Pittapapda)
5a. Plants with thorns or spines 6
5b. Plants without thorns or spines 8
6a. Flowers yellow .... *Barleria prionitis* (Kurantak)
6b. Flowers blue, purple or pink 7
7a. Plants growing along sea-shore, leaf-margin spinous .... *Acanthus ilicifolius* (Marandi)
7b. Plants growing in road-side ditches or in wet localities, leaf margin not spinous .... *Hygrophila schulli* (Kokilaksha)
8a. Flowers white or with bluish or pinkish ting 9

8b. Flowers red, blue, purple or orange coloured 11

9a. Shrubs, 2-3 meters or more tall, flowers in dense spikes .... *Adhatoda zeylanica* (Vasa)

9b. Undershrubs, up-to 1 meter tall; flowers solitary or in lax cymes or racemes 10

10a. Flowers solitary or 2-3 together; corolla 2-3 cm long; capsules 4- seeded, contracted below into a narrow, long stalk.................. *Rhinacanthus nasutus* (Yuthik Parni)

10b. Flowers in lax racemes; corolla up-to 1 cm. long; capsules many-seeded, without a narrow stalk .... *Andrographis paniculata* (Kalamegh)

11a. Flowers blue or purple, solitary or in lax axillary spikes .... *Barleria cristata* (Neel Kurantak)

11b. Flowers blue but turning red, pink or orange colored, in dense terminal spikes.. 12

12a. Leaves opposite; flowers blue but turning red or pink .... *Eranthemum roseum* (Dashmul)

12b. Leaves in pseudowhorls, flowers red or orange colored .... *Crossandra infundibuliformis* (Aboli.

2. FAMILY - Adiantaceae (HANSPADI Kul.)

1a. Fronds (Leaves) pinnate 2

1b. Fronds pedate .... *Adiantum pedatum* (Hansraj.

2a. Fronds unipinnate; leaflets broader than long ....*Adiantum philippense* (Hanspadi)

2b. Fronds bi- or tripinnate; leaflets longer than broad .... *Adiantum capillus-veneris* (Hansraj)

3. FAMILY- Agaricaceae (CHHATRAK Kul.)

1a. Plants growing on ground, umbrella shaped .... *Psalliota campestris* (Chhatrak)

1b. Plants growing on dead wood, bracket shaped ....*Psalliota ostreata* (Phanas Alambe)

4. FAMILY - Aizoaceae (VASU Kul.)

Trianthema portulacastrum (Syn. *T. monogyna*) (Vasu)

5. FAMILY - Alangiaceae (ANKOL Kul.)

Alangium salviifolium (Ankol)
(Syn. *A. lamarckii*)

6. FAMILY- Alliaceae (PALANDU Kul.)

1a. Underground bulb simple, leaves cylindrical, hollow…… ....*Allium cepa* (Palandu)

1b. Underground bulb made-up of many scales, leaves flat….. ....*Allium sativum* (Lason)

7. FAMILY – Altingiaceae

Liquidambar orientalis

8. FAMILY – Amaranthaceae (APAMARGA Kul)

1a. Leaves opposite; flowers bisexual ....*Achyranthes aspera* (Apamarga, Aghada)

1b. Leaves alternate; flowers bisexual or unisexual 2

2a. Seed one 3

2b. Seeds 2-10 .... *Celosia argentea* (Shitwar, Kurdu)

3a. Stem and leaves densely covered with ash- colored or gray hairs ....*Aerva lanata* (Shatakabhedi, Kapurphuti)

3b. Stem and leaves not as above 4

4a. Plants with thorns ....*Amaranthus spinosus* (Tanduliya, Kate Math)

4b. Plants without thorns 5

5a. Flowers in terminl paniculate spikes ....*Amaranthus paniculatus* (Chuamars, Rajgira)

5b. Flowers in terminal spikes or axillary clusters 6

6a. Flowers in axillary clusters only .... *Amaranthus roxburghiahus* (Marish, Chauli)

6b. Flowers in axillary clusters as well as in terminal spikes................................ *Amaranthus tricolor* (Marish, Chauli)

9. FAMILY - Amaryllidaceae (NAGDAUNA Kul.)

1a. Leaves 10-20 cm wide .... *Crinum asiaticum* (Nagdauna)

1b. Leaves up-to 10 cm wide .... *Crinum latifolium* (Gadanicha Kanda)

10. FAMILY - Anacardiaceae (AMRA Kul.)

1a. Leaves simple 2

1b. Leaves compound 5

2a. Flowers polygamous 3

2b. Flowers bisexual .... *Buchanania lanzan* (Char, Charoli)

3a. Stamens 1-5; fruit a fleshy drupe with slender peduncle ....*Mangifera indica* (Amra)

3b. Stamens 5-10; fruit a nut with fleshy fruit like peduncle 4

4a. Stamens 5-6; styles 3 ....*Semecarpus anacardium* (Bhallatak)

4b. Stamens 8-10; style 1 ....*Anacardium occidentale* (Kaju)

5a. Shrubs with 1-5-foliolate leave. ...flets crenate margined ....*Rhus parviflora* (Sumak)

5b. Tree with pinnate leaves; leaflets e... margined ....*Lannea coromandelica* (Moi, Shimpti)

11. FAMILY - Annonace... (SITAPHAL Kul.)

1a. Ovaries united in fruit 2

1b. Ovaries distinct in fruit .... *Polyalthia longifolia* (Panacha Ashok)

2a. Fruits green when ripe; epicarp studded with raised areoles .... *Annona squamosa* (Sitaphal)

2b. Fruits brick-red when ripe; epicarp smooth .... *Annona reticulata* (Ramphal)

12. FAMALY – Apiaceae (JIRAK Kul.)

1a. Plants trailing with thread like slender running stem; leaves reniform; flowers in simple umbels .... *Centella asiatica* (Manduk Parni)

1b. Plants erect with normal stem; leaves various, often divided into many narrow segments; flowers in compound umbels 2

2a. Ultimate leaf- segments very narrow 3

2b. Ultimate leaf- segment broad 6

3a. Flowers white or pink; fruit bristly .... *Cuminum cyminum* (Jirak)

3b. Flowers yellow; fruit not bristly 4

4a. Rays equal or nearly so; umbels clustered at tip of the stem; the central umbel nearly sessile; fruit 1-1.5 cm long .... *Ferula asafoetida* (Hingu)

4b. Rays unequal; umbles not as above; fruits not more than 6 mm long 5

5a. Fruit flattened with distinct wing .... *Anethum graveolens* (Mishreya)

5b. Fruit not flattened, not winged .... *Foeniculum vulgare* (Madhurika)

6a. Stems and leaves hairy .... *Daucus carota* (Garjar)

6b. Stems and leaves glabrous 7

7a. Outer flowers of umbellets with enlarged ray like petals .... *Coriandrum satiuum* (Dhanyak)

7b. Outer flowers not as above 8

8a. Leaves alternate .... *Angelica glauca* (Chorak)

8b. Leaves pinnate 9

9a. Involucral bracts lacking or minute 10

9b. Involucral bracts conspicuous .... *Apium graveolens* (Ajmod)

10a. Fruit compressed .... *Carum carvi* (Sushapi, Shahajire)

10b. Fruit not compressed .... *Trachyspermum ammi* (Yavani Onva)

13. FAMILY- Apocynaceae (KUTAJ Kul.)

1a. Trees or shrubs 2

1b. Herbs or undershrubs 10

2a. Leaves alternate; 3

2b. Leaves opposite or whorled; 4

3a. Flowers yellow or orange coloured .... *Thevetia peruviana* (Peet Karveer)

3b. Flowers white or reddish .... *Plumeria rubra* (Kshirchampak)

4a. Leaves opposite 5

4b. Leaves whorled 9

5a. Shrubs armed with thorns 6

5b. Shrubs or trees, without thorns 7

6a. Corolla 2 cm. or more long;
fruits 1.5 cm. long .... *Carissa congesta* (Karmardak, Karvand)

6b. Corolla less than 2 cm. long; fruit 6-8 mm long .... *Carissa opaca* (Karmardak, Karvand)

7a. Scandent or twining shrubs .... *Ichnocarpus fruitescens* (Shyamlata)

7b. Erect shrubs or trees 8

8a. Cultivated plants; corolla tube very narrow.. *Taberenaemontana coronaria* (Tagar)

8b. Wild plants, corolla tube not very narrow .... *Holarrhena pubescens* (Kutaj)

9a. Trees with 5-9 leaves in a whorl; flowers
less than 1 cm. long .... *Alstonia scholaris* (Saptaparmi)

9b. Shrubs with 3 leaves in a whrol; flowers
3 cm or more long.. *Nerium oleander* (Karveer)

10a. Leaves whorled; fruits drupaceous .... *Rauwulfia serpentina* (Sarpagadha)

10b. Leaves opposite, fruit follicular.. *Catharanthus roseus* (Sadaphuli)

14. FAMILY – Araceae (SHURAN Kul.)

1a. Leaf solitary, divided into many segments 2

1b. Leaves 2- many, entire 4

2a. Leaves bulbiliferous .... *Amorphophallus bulbifer* (Shuran)

2b. Leaves not bulbiliferous 3

3a. Spadix as long as the spathe .... *Amorphophallus paeoniifolius* (Shuran)

3b. Spadix longer than the spathe .... *Amorphophallus sylvaticus* (Aranya Shuran)

4a. Rhizome long creeping; leaves linear,
without petiole; flowers bisexual with
trimerous perianth .... *Acorus calamus* (Vacha)

4b. Rhizome short; leaves ovate with long
petiole; flowers unisexual, without perianth 5

5a. Petiole attached in the centre of the blade
(leaves peltate); ovules many on parietal
placentas .... *Colocasia esculenta* (Kachvi, Aalu)

5b. Petiole attached at the base of the blade;
ovules few on basal placenta .... *Alocasia indica* (Kasalu)

15. FAMILY – Arecaceae (NARIKEL Kul.)

1a. Leaves long, pinnatisect 2

1b. Leaves fan- shaped .... *Borassus flabellifer* (Tal, Tad)
2a. Lower leaf- segments turning into spines 3
2b. Lower leaf- segments not turning into spines 4
3a. Stems many from the ground; fruits with thick sweet flesh .... *Phoenix dactylifera* (Kharjuri)
3b. Stem solitary, very rarely branching near the top; fruits with thin epicarp .... *Phoenix sylvestris* (Vit Kharjuri)
4a. Leaf segments short, broadening and jagged towards the top .... *Caryota urens* (Surmad)
4b. Leaf segments long, narrowed at the fine point towards the top 5
5a. Stem scandent; fruits covered with hard scales........6
5b. Stems erect; fruits not as above 7
6a. Spathes tubular, persistent........ *Calamus draco* (Hira Dakhan)
6b. Spathes open, deciduous........ *Daemonorhops kurzianus*
7a. Fruit appearing like a double coconut .... *Lodoicea maldivica* (Daaryacha Naral)
7b. Fruit not as above........8
8a. Trunk columnar; fruits 4-5 cm long .... *Areca cathecu* (Pug. Pophal)
8b. Trunk not columnar; fruits 20-30 cm long .... *Cocos nucifera* (Narikel)

16. FAMILY – Aristolochiaceae (ISHWARI Kul.)

1a. Plants 20-60 cm long; leaves ovate, ash-coloured .... *Aristolochia bracteolata* (Dhumrapatra, Gindhan)
1b. Plants more than 60 cm long; leaves oblanceolate, green .... *Aristolochia indica* (Ishwari, Sapsan)

17. FAMILY – Asclepiadaceae (ARK Kul.)

1a. Plants erect 2
1b. Plants twining or scandent 3
2a. Petals erect, ovate, purple tipped; corona lobes longer than the staminal column .... *Calotropis procera* (Ark, Rui)
2b. Petals spreading, narrow, white; corona lobes shorter than the staminal column .... *Calotropis gigantea* (Ark, Rui)
3a. Leafless shrubs .... *Sarcostemma acidum* (Ransher)
3b. Leafy shrubs 4
4a. Anthers with a membraneous appendage 5
4b. Anthers without a membraneous appendage 8
5a. Flowers rather large, purplish, pollinia pendulous .... *Holostemma ada - kodien* (Shirdodi)

5b. Flowers small, greenish-white; pollinia erect 6

6a. Corona attached to corolla .... *Gymnema sylvestre* (Mesh shrungi, Gudmar)

6b. Corona attached to stamens 7

7a. Corona lobes more or less free .... *Tylophora indica* (Pittwalli)

7b. Corona lobes with their bases attached to staminal column....................*Marsdenia tenacissima*

8a. Twining shrubs; corolla with spreading petals and a short tube .... *Leptadenia reticulata* (Jivanti)

8b. Twining herbs; corolla with the petals connate at the tips and a long tube ....*Ceropegia bulbosa* (Patal Tumbadi)

18. FAMILY – Asteraceae (SHATAPATRI Kul.)

1a. Plants spinous 2

1b. Plants not spinous 3

2a. Stem and leaves beneath white hairy; florets white....................*Echinops echinatus* (Utkantak)

2b. Stem and leaves beneath not white hairy; florets yellow or orange ....*Carthamus tinctorius* (Kusumbh)

3a. Pappus of bristles or hairs, not widened at base....................................... 4

3b. Pappus of scales or awns lacking 14

4a. Ray or ligulate florets lacking 5

4b. Ray or ligulate florets present .... .11

5a. Leaves opposite .... *Eupatorium ayapana* (Ayapana)

5b. Leaves alternate or basal 6

6a. Corollas purple 7

6b. Corollas yellow 10

7a. Leaves basal only; flower heads subtended by leafy bracts .... *Elephantopus scaber* (Gojivha)

7b. Leaves alternate or basal as well as alternate; heads not subtended by leafy bracts....................................... 8

8a. Involucral bracts in many series....................................... 9

8b. Involucral bracts in one series .... *Emelia sonchifolla* (Sadmandi)

9a. Involucral bracts foliaceous; heads 1-3 cm or more across; pappus dull brown .... *Centrantherum anthelminticum* (Aranya Jirak)

9b. Involucral bracts not foliaceous; heads less than 1 cm across; pappus white ....*Vernonia cinerea* (Sahdevi)

10a. Involucral bracts dry .... *Pluchea lanceolata* (Patrarasna)

10b. Involucral bracts green ....*Blumea lacera* (Kukundar)

11a. All florets with ligulate corollas 12

11b. Outer florets in the head with ligulate corollas, inner florets with tubular corollas 13

12a. Leaves all basal, some of them pinnatifid; plants of high altitudes .... *Taraxacum officinale* (Dugdhapheni)
12b. Leaves basal as well as on the stem; pinnatifid or not; plants of lower altitudes .... *Launaea procumbens* (Pathari)
13a. Leaves entire .... *Inula racemosa* (Rasna)
13b. Leaves pinnatifid .... *Anacyclus pyrethrum* (Aakarkarabh)
14a. Rays lacking or inconspicuous 15
14b. Rays present 21
15a. Flower head solitary 16
15b. Flower head in racemes or panicles 20
16a. Flower head yellow 17
16b. Flower heads white, bluish or purple 19
17a. Plants prostrate or spreading; heads globose 18
17b. Plants erect; heads ovoid .... *Spilanthes oleracea* (Khota Akkalkara)
18a. Leaves entire or toothed .... *Centipeda minima* (Chhikkika)
18b. Leaves pinnatifid .... *Grangea maderaspatana* (Machipatri)
19a. Plants of high altitudes; stems 1-2 m tall; heads 2.5-3.5 cm in diam .... *Saussurea lappa* (Kushtha, Uplet)
19b. Plants of low altitudes; stems upto 60 cm. tall; heads 1-1.5 cm. in diam .... *Sphaeranthus indicus* (Gorakhmundi)
20a. All florets in the head similar .... *Artemisia maritima* (Chauhar, Kirmani ajwan)
20b. Outer florets female, central florets bisexual .... *Artemisia nilagirica* (Surparn, Surpin)
21a. Leaves at least the lower, opposite 22
21b. All leaves alternate 26
22a. Lower leaves opposite; the upper alternate; involucral bracts 1-seriate; pappus absent .... *Guizotia abyssinica* (Ramtil)
22b. All leaves opposite; involucral bracts 1-many-seriate; pappus present 23
23a. Involucral bracts united into a cup or tube; receptacle not paleaceous .... *Tagetes erecta* (Zendu)
23b. Involucral bracts not so united; receptacle paleaceous........ 24
24a. Palea of the receptacle concave embracing the cetral florets; rays yellow .... *Wedelia chinensis* (Peet Bhrungraj)
24b. Palea of the receptacle flat, not embracing the central florets; rays white, rarely with yellow tinge 25
25a. Leaves entire; central florets with white corollas; pappus absent or represented by two awns .... *Eclipta prostrata* (Bhrungraj)

25b. Leaves pinnatifid or lobed; central florets with
yellow corollas; pappus-scales plumose .... *Tridax procumbens* (Dagadi Pala)

26a. Leaves dissected into fine segments 27

26b. Leaves not as above 28

27a. Head 2-2.5 cm. across; rays 10-20, white,
emarginate; achene 3-5 ribbed; eciliate;
pappus absent .... *Matricaria chamomilla* (Babuna)

27b. Heads 5-8 mm. across; ray often only one;
yellow; achenes compressed, ciliate along margins;
pappus of two stiff, spreading awns .... *Glossocardia bosvallea* (Khadak shepu)

28a. All florets with blue, purple or white
ligulate corollas; achenes long beaked .... *Cichorium intybus* (Kasni, Chikori)

28b. Marginal or rarely all florets with yellow or
white ligulate corollas; achenes not beaked .... *Dendrathema indica* (Shatapatri)

19. FAMILY - Averrhoaceae (KARMARANG Kul.)

Averrhoa carambola (Karambola, Karmaranga)

20. FAMILY - Balanitaceae (HINGAN Kul.)

Balanites aegyptiaca (Hingan)

21. FAMILY - Bambusaceae (VANSH Kul.)

Bambusa arundinacea (Vansh, Velu)

22. FAMILY - Berberidaceae (DARUHARIDRA Kul.)

1a. Plants shrubby; anthers opening by valves .... *Berberis aristata* (Daruharidra)

1b. Plants herbaceous; anthers not opening by valves.......................... *Podophyllum emodi* (Giriparpat)

23. FAMILY - Betulaceae (Corylaceae) (BHURJPATRA Kul.)

Betula alnoides (Bhujpattra)

24. FAMILY - Bignoniaceae (ARALU Kul.)

1a. Leaves simple; flowers orange-yellow .... *Tecomella undulata* (Rohit)

1b. Leaves compound; flowers not orange-yellow 2

2a. Leaves unipinnate; capsules spirally twisted 3

2b. Leaves 3-pinnate; capsules not spirally twisted .... *Oroxylum indicum* (Shyonak, Aralu)

3a. Calyx spathaceous, capsules split down
on one side .... *Dolichandrone falcata* (Meshshrungi)

3b. Calyx campanulate, 2-5 lobed 4

4a. Inflorescence glandular pubescent .... *Stereospermum chelenoides* (Patla)

4b. Inflorescence not glandular pubescent .... *Stereospermum colais* (Shwet Patla)

25. FAMILY - Bixaceae (SINDURI Kul.)

Bixa orellana (Sinduri)

26. FAMILY – Bombacaceae (SALMALI Kul.)

1a. Stem and branches green; flowers pale pink; stamens 5. .... *Ceiba pentandra* (Kut Shalmali)

1b. Stem and branches not green; flowers white or red; stamens many 2

2a. Stem and branches often prickly; flowers red; stamens in 5-6 bundles .... *Bombax ceiba* (Salmali)

2b. Stem and branches not prickly; flowers white; stamens monadelphous .... *Adansonia digitata* (Ghorakh amli)

27. FAMILY – Boraginaceae (BHURUNDI Kul.)

1a. Leaves opposite, sub-opposite or the upper most alternate, clasping the stem at their base; stamens exserted from the corolla tube .... *Trichodesma indicum* (Jinghini)

1b. Leaves all alternate, not clasping the stem; stamens included within the corolla tube 2

2a. Stamens deeply sagittate; ovary deeply divided into 4 parts .... *Onosma bracteatum* (Nalika)

2b. Stamens not deeply sagittate; ovary not deeply divided .... *Heliotropium indicum* (Bhurundi)

28. FAMILY – Brassicaceae (RAJIKA Kul.)

1a. Fruits long, many seeded 2

1b. Fruits short, few seeded .... *Lepidium sativum* (Chandrashur)

2a. Fruits spongy, indehiscent .... *Raphanus sativus* (Mulak)

2b. Fruits not spongy, dehiscent 3

3a. Fruits dehiscing throughout .... *Sisymbrium irio* (Khakasi)

3b. Fruits with indehiscent 1-seeded beak 4

4a. Leaves thick; flowers pale yellow, 6-20 mm across; sepals not saccate.......................... *Brassica napus*

4b. Leaves thin; flowers bright yellow, 5-10 mm across; sepals saccate 5

5a. Fruits 3-5 cm long, many seeded 6

5b. Fruits 2-3 cm long; 5-10 seeded .... *Brassica nigra* (Sarsap)

6a. Seeds red .... *Brassica juncea* (Rajika)

6b. Seeds white . ... *Brassica alba* (Shwet Rajika)

29. FAMILY – Bromeliaceae (ANANAS Kul.)

Ananas comosus (Anannas)

30. FAMILY – Burseraceae (GUGGULU Kul.)

1a. Bark papery; flowers bisexual, white; disc annular, crenate; fruits dehiscent .... *Boswellia serrata* (Sallaki)

1b. Bark not papery, smooth; flowers polygamous; reddish-yellow; disc cupular; fruits indehiscent.......... 2

2a. Leaves 1-3 foliolate, side leaflets often very small or absent .... *Commiphora wightii* (Guggulu)

2b. Leaves pinnate .... *Commiphora myrrha* (Hirabol)

31. FAMILY - Cactaceae (KANTHARI Kul.)

1a. Plants with small glochids as well as long spines.......... *Opuntia elatior* (Kantseb)

1b. Plants with glochids but without spines .... *Nopalea cochenillifera* (Nivdung.

32. FAMILY – Caesalpiniaceae (PUTIKARANJ Kul.)

1a. Leaves 1-foliolate and lobed 2

1b. Leaves pinnately compound 3

2a. Leaves broader than long, 2.5 x 2.5 - 6 cm; calyx spathaceous; corolla white or pale yellow, 1-1.5 cm long .... *Bauhinia racemosa* (Ashmantak)

2b. Leaves longer than broad, 10-12 x 8-10 cm; calyx unequally 5-toothed; corolla pinkish purple; one petal variegated .... *Bauhinia variegata* (Kanchnar)

3a. Leaves unipinnate 4

3b. Leaves 2-3 pinnate 13

4a. Petals present 5

4b. Calyx oranged or red coloured; petals absent .... *Saraca asoca* (Ashok)

5a. Petals 5, dull yellow with red lines; stamens 3, filaments connate at base .... *Tamarindus indica* (Aamlika)

5b. Petals 5, variously coloured; stamens 3-10, filaments not connate 6

6a. Tree with golden yellow flowers in drooping racemes; pods cylindric, 30-60 cm long .... *Cassia fistula* (Aaragwadh)

6b. Shrubs or undershrubs; flowers yellow but not in drooping racemes; pods flat or cylindric, often less than 20 cm long 7

7a. Undershrubs with 2-3 pairs of obtuse leaflets 8

7b. Shrubs with 3-10 or more pairs of acute leaflets 10

8a. Leaflets 2 pairs; flowers brownish- yellow;
pods flat, 2.5-4 cm long .... *Cassia absus* (Vanya kulathya, Chinol)

8b. Leaflets 3 pairs; flowers yellow;
pods cylindric or 4- gonous, 10-20 cm long 9

9a. Leaves with one gland on the rachis;
calyx hairy; anthers of the three longest
stamens narrowed at apex; pods
often 4-gonous .... *Cassia obtusifolia* (Chakramard)

9b. Leaves with two glands on the rachis;
calyx glabrous or nearly so; anthers of the
three longest stamens not narrowed at apex;
pods cylindric .... *Cassia tora* (Chakramard)

10a. Leaflets 3-5 pairs, acute or acuminate at apex .... *Cassia occidentalis* (Kasmard)

10b. Leaflets 5-12 pairs, subacute,
obtuse or apiculate at apex 11

11a. Stipules foliaceous, auricled, persistent; .... *Cassia auriculata* (Aawartaki)

11b. Stipules linear- subulate, deciduous 12

12a. Plants wild; leaflets sub-acute at apex .... *Cassia sophera* (Kasmard)

12b. Plants cultivated; leaflets apiculate at apex .... *Cassia angustifolia* (Sonamukhi)

13a. Scandent shrubs often densely prickly 14

13b. Erect shrubs or trees, rather
smooth or laxly prickly 16

14a. Flowers sessile in simple or
branched slender spikes .... *Maullava spicata* (Gucha karanj)

14b. Flowers pedicelled in simple or
branched racemes 15

15a. Flowers greenish- yellow; pods prickly .... *Caesalpinia bonduc* (Putikaranj)

15b. Flowers golden yellow; pods smooth .... *Caesalpinia decapetala* (Chilar)

16a. Flowers orange or yellow; filaments
glabrous .... *Caesalpinia pulcherrima*
(Krishnachuda, Sankeshwar)

16b. Flowers yellow; filaments densely wooly at base .... *Caesalpinia sappan* (Patang)

33. FAMILY- Cannabaceae (GAANJA Kul.)

Cannabis sativa (Ganja)

34. FAMILY – Capparaceae (WARUN Kul.)

1a. Leaves 3- foliolate; sepals inserted on the edge of the disc .... *Crateva adansonii* (Warun)

1b. Leaves simple; sepals inserted on the torus 2

2a. Stamens 4-6, inserted on gynophore; disc
produced on one side like a tube .... *Cadaba fruticosa* (Habab)

2b. Stamens 8-20, inserted on the torus; disc absent or not as above 3
3a. Leaves early deciduous; sepals and petals red.... *Capparis decidua* (Karir)
3b. Leaves persistent; sepals and petals white or dull yellow 4
4a. Branchlets and young leaves densely purplish-grey tomentose; filamens turning pinkish-purple.... *Capparis zeylanica* (Vyaghranakhi)
4b. Branchlets and young leaves glabrous or nearly so; filaments not turning pinkish – purple .... *Capparis sepiaria* (Kakadani)

35. FAMILY - Caricaceae (MADHUKARKATI Kul.)

Carica papaya (Madhukarkati, Popai)

36. FAMILY - Cassythaceae (AAKASHVEL Kul.)

Cassytha filiformis (Aakashvel)

37. FAMILY - Celastraceae (JYOTISHMATI Kul.)

1a. Erect shrubs or small trees .... *Cassine albens* (Budhkeshi)
1b. Scandent shrubs 2
2a. Plants thorny; flowers in short panicles .... *Maytenus emarginata* (Henkal)
2b. Plants not thorny; flowers in long racemes .... *Celastrus paniculata* (Jyotishmati)

38. FAMILY - Ceratophyllaceae (SHAIWAL Kul.)

Ceratophyllum demersum (Shaiwal)

39. FAMILY - Chenopodiaceae (PALAKYA Kul.)

1a. Plants twining .... *Basella rubra* (Upodaki)
1b. Plants not twining 2
2a. Root swollen, napiform, red inside .... *Beta vulgaris* (Raktagrunjan)
2b. Root not as above 3
3a. Flowers unisexual; perianth of two kinds .... *Spinacia oleracea* (Palak)
3b. Flowers bisexual; perianth of one kind 4
4a. Plants glandular- pubescent, strong smelling .... *Chenopodium ambrosioides* (Chandil)
4b. Plants not glandular- pubescent, not strong smelling.... 5
5a. Tall herb; seeds smooth .... *Chenopodium album* (Chilli)
5b. Small herb; seeds rugose .... *Chenopodium murale*

40. FAMILY - Cleomaceae (TILWAN Kul.)

1a. Plants growing in or near water; flowers pink .... *Cleome chelidonii*
1b. Plants growing on dry land; flowers not pink 2
2a. Flowers white; stamens raised on androphore .... *Cleome gynandra* (Tilparni)
2b. Flowers yellow; stamens hypogynous .... *Cleome viscosa* (Adityabhakta)

41. FAMILY – Clusiaceae (SARAMLA Kul.)

1a. Style absent; stigma sessile or subsessile 2
1b. Style present stigma on slender style 6
2a. Calyx with 4-5 sepals 3
2b. Calyx closed in bud, opening into two valves .... *Mammea longifolia* (Nameru, Surangi)
3a. Stamens of the male flowers in 4 angled mass; anthers 4- angled 4
3b. Stamens of the male flowers in globole mass; anthers orbicular. .... *Garcinia morella* (Tamal)
4a. Fruits with 7-12 longitudinal furrows, mamillate at the tip .... *Garcinia cambogia* (Saramla)
4b. Fruit smooth; not mamillate but depressed at the tip 5
5a. Fruits 3-5 cm. in diameter .... *Garcinia indica* (Saramla)
5b. Fruits 15-20 cm in diameter .... *Garcinia pedunculata* (Amlavetas, Thaikal)
6a. Leaves shining green with parallel veins; flowers up-to 2 cm across, in racemes or panicles .... *Calophyllum inophyllum* (Punnag)
6b. Leaves dull green; flowers 4-8 cm across solitary, axillary .... *Mesua nagassaricum* (Nagshekhar)

42. FAMILY - Cochlospermaceae (SUVARNA KARPAS Kul.)

Cochlospermum religiosum (Suvarnakarpas)

43. FAMILY - Combretaceae (HARITAKI Kul.)

1a. Trees; petals absent 2
1b. Scandent shrubs; petals 4-5 6
2a. Flowers in dense globose heads .... *Anogeissus latifolia* (Dhav, Dhawada)
2b. Flowers in simple or branched spikes 3
3a. Fruits not winged 4
3b. Fruits with broad hard wings 5
4a. Leaves abtuse or rounded at apex; flowers in simple spikes; fruits hairy, not ribbed .... *Terminalia bellirica* (Bibhit)

4b. Leaves acute at apex; flowers in branched spikes; fruits with 4-5 conspicuous ribs .... *Terminalia chebula* (Haritak)

5a. Bark smooth, dull white; veins of the fruit wings curving upwards .... *Terminalia arjuna* (Arjuna)

5b. Bark rough, blackish; veins of the fruit wings straight, horizontal .... *Terminalia alata* (Asan)

6a. Wild plants; flowers small, white or cream coloured in dense spikes; fruits winged .... *Combretum albidum*

6b. Cultivated plants; flowers large, white turning red in pendulous corymbs; fruits not developing .... *Quisqualis indica* (Rangunvel)

44. FAMILY – Convolvulaceae (NISHOTTAR Kul.)

1a. Plants erect 2

1b. Plants twining, trailing or creeping 4

2a. Plants upto 15 cm tall; flowers sessile in terminal clusters; styles 2 .... *Cressa cretica* (Rudrawanti)

2b. Plants more than 15 cm tall; flowers pedicelled, axillary solitary; style one often 2-fid........................ 3

3a. Flowers rose-yellow coloured .... *Convolvulus microphyllus* (Shankhpushpi)

3b. Flowers white .... *Convolvulus pluricaulis* (Shankhpushpi)

4a. Stem and branches with broad membranous wings........................ *Operculina turpethum* (Nishottar)

4b. Stem and branches not as above 5

5a. Plants growing in water .... *Ipomoea aquatica* (Kalambi, Naal)

5b. Plants growing on land 6

6a. Plants growing on sand along sea- shore; leaves 2-lobed........................ *Ipomoea pes-caprae* (Maryada vel)

6b. Plants growing inland; leaves not 2 lobed 7

7a. Plants prostrate, trailing or creeping on the ground; stem slender 8

7b. Plants twining; stem rather stout 9

8a. Leaves elliptic-lanceolate; flowers blue .... *Evolvulus alsinoides* (Vishnukranta, Shankhavali)

8b. Leaves ovoid-reniform; flowers yellow .... *Merremia emarginata* (Mushak parni)

9a. Fruit hard, indehiscent or breaking irregularly 10

9b. Fruits with thin wall, dehiscent, often 3-4-valved 11

10a. Stem and leaves beneath covered with dense silvery-white hairs; flowers rose-coloured in sub-capitate cymes .... *Argyreia nervosa* (Samudrashok)

10b. Stems and leaves beneath white pubescent, flowers dull-white in axillary 1-7 flowered lax cymes .... *Rivea laotica* (Fanjika)

11a. Leaves entire, not lobed 12

11b. Leaves 3-5 or more lobed 13

12a. Flowers white with 10-15 cm long, narrow corolla tube .... *Ipomoea alba* (Chandrakant)

12b. Flowers pink or rose-coloured with 2-3 cm long corolla tube .... *Ipomoea sepiaria*(Lakshmana)

13a. Plants cultivated for root tubers .... *Ipomoea batatas* (Ratale)

13b. Plants wild 14

14a. Roots not tuberous; leaves 3-lobed 15

14b. Roots large tuberous; leaves 3-9-lobed .... *Ipomoea mauritiana* (Kshirvidari Kanda)

15a. Stem, leaves and sepals hairy; corolla blue; ovary 3-locular, 6-seeded .... *Ipomoea nil* (Shankhini)

15b. Stem, leaves and sepals glabrous; corolla white or pink; ovary 4-locular, 4-seeded .... *Convolvulus arvensis* (Hinranpag)

45. FAMILY - Crassulaceae (PARNABEEJ Kul.)

1a. Leaves simple as well as 3-5-foliolate on the same plant; flowers pink .... *Kalanchoe pinnata* (Parnabeej)

1b. All leaves 3-7 foliolate; flowers yellow .... *Kalanchoe laciniata* (Hemsagar)

46. FAMILY - Cucurbitaceae (KARKATI Kul.)

1a. Leaves 3-7- lobed; lobes pinnatifid; segments abuse or rounded at apex 2

1b. Leaves 3-7-lobed; lobes entire or toothed; segments acute 3

2a. Plants culitivated; fruits globose or ellipsoid, 15-30 cm in diameter; filled with red juicy edible pulp .... *Citrullus lanatus* (Kaling)

2b. Plants wild; fruits globose, 5-8 cm in diameter; filled with dry, fibrous, very bitter pulp. .... *Citrullus vulgaris* (Indravaruni)

3a. Flowers white 4

3b. Flowers greenish-yellow or yellow 9

4a. Petals fimbriate with thread-like processes 5

4b. Petals not fimbriate 8

5a. Fruits globose .... *Trichosanthes tricuspidata* (Mahakas)

5b. Fruits elongate 6

6a. Fruits more than 30 cm long .... *Trichosanthes anguina* (Padval)

6b. Fruits upto 15 cm long 7

7a. Wild plants; fruits with narrow pointed apex ....*Trichosanthes cucumerina* (Ranpadval)

7b. Cultivated plants; fruits with acute or rounded apex ....*Trichosanthes dioica* (Patol)

8a. Plants hairy; leaves 10-25 cm wide; corolla tube long and narrow; fruits club-shaped, 15-90 cm. long ....*Lagenaria siceraria* (Katutumbi)

8b. Plants glabrous; leaves 5-15 cm wide; corolla tube short and broad; fruits ellipsoid, 5-12 cm long. ....*Coccinia grandis* (Bimbi)

9a. Corolla gamopetalous, shallowy lobed 10

9b. Corolla nearly polypetalous, deeply lobed 15

10a. Anther-cells flexuose or conduplicate; fruits large 11

10b. Anther cells straight; fruits small 13

11a. Petioles with rigid and pungent hairs; fruit surface rough ....*Cucurbita pepo* (Dangri)

11b. Petioles not with hairs as above; fruit surface smooth 12

12a. Calyx lobes narrow, subulate; fruits reddish brown ....*Cucurbita maxima* (Lal Bhopla)

12b. Calyx lobes broad, spathulate, leafy; fruits greenish, blotched with white ....*Cucurbita moschata* (Lal Bhopla)

13a. Plants rough; flowers yellow; fruits globose, 5-6 mm in diameter 14

13b. Plants glabrous; flowers greenish- yellow; fruits elongate, 2-3 cm long ....*Corallocarpus epigaeus* (Katunahi)

14a. Plants monoecious; flowers bright yellow ....*Melothria maderaspatana* (Ghantali)

14b. Plants dioecious; flowers pale yellow ....*Melothria heterophylla* (Gometi)

15a. Plants more or less glabrous 16

15b. Plants hairy 22

16a. Wild plants; flowers pale yellow; fruits globose, dark- red with vertical lines ....*Diplocyclos palmata* (Baj, Shivlingi)

16b. Cultivated plants; flowers bright yellow; fruits elongate, not as above 17

17a. Male flowers with large enveloping bract; fruits fleshy, red when ripe, echinate or muricate 18

17b. Male flowers not as above; fruits dry, brown when ripe, smooth echinate or ribbed 19

18a. Plants wild, dioecious; fruits echinate, 5-6 x 2-3 cm ....*Momordica dioica* (Karkotaki)

18b. Plants cultivated, monoecious; fruits muricate, 5-30 x 3-7 cm ....*Momordica charantia* (Karwell)

19a. Plants wild; fruits upto 15 cm long 20
19b. Plants cultivated, fruits more than 15 cm long 21
20a. Fruits ribbed, 5-10 cm long .... *Luffa amara* (Koshataki)
20b. Fruits echinate, 2-4 cm long .... *Luffa echinata* (Devdali)
21a. Fruits ribbed .... *Luffa acutangula* (Mahakoshataki)
21b. Fruits not ribbed .... *Luffa cylindrica* (Rajkoshataki)
22a. Filaments of stamens exserted from the calyx tube; fruits 25-60 x 10-25 cm, densely hairy when young, thickly waxy when mature .... *Benincasa hispida* (Kushmand)
22b. Filaments of stamens included within the calyx tube; fruits elongate, not as above 23
23a. Plants wild; fruits 3-10 x 2.5-5 cm, pale yellow with green longitudinal stripes, bitter .... *Cucumis callosus* (Karit)
23b. Plants cultivated; fruits more than 10 cm long or globose 24
24a. Fruits elongate 25
24b. Fruits globose or ellipsoid .... *Cucumis melo* (Kharbuj)
25a. Fruits 10-15 cm long, prickly .... *Cucumis stativus* (Trapus)
25b. Fruits more than 15 cm long smooth .... *Cucumis utilissimus* (Karkati)

47. FAMILY - Cupressaceae (HOUBER Kul.)

Juniperus communis (Houber)

48. FAMILY - Cyperaceae (NAGARMUSTAK Kul.)

1a. Flowers in dense white heads .... *Kyllinga bulbosa* (Nirvishi)
1b. Flowers in lax umbels or panicles 2
1b. Rhizome stout, long, stems 2-3 m tall; flowers in panicles .... *Scirpus kysoor* (Kachra)
2a. Stems slender; leaves narrow; root tubers, strong smelling .... *Cyperus scariosus* (Nagar Mustak)
2b. Stems stout; leaves broad; root tubers, mild smelling .... *Cyperus rotundus* (Mustak)

49. FAMILY – Dilleniaceae (BHAVYA Kul.)

1a. Flowers 10-15 cm across .... *Dillenia indica* (Bhavya)
1b. Flowers 2-3 cm across .... *Dillenia pentagyna*

50. FAMILY – Dioscoreaceae (KARANDA Kul.)

1a. Plants with bulbils in the leaf axils 2
1b. Plants without bulbils in the leaf axils 3
2a. Stem terete; bulbils warted .... *Dioscorea bulbifera* (Kadu Karanda)
2b. Stem angled or winged, bulbils smooth .... *Dioscorea alata* (Dandalu)

3a. Root tubers spinous, edible .... *Dioscorea esculenta* (Kange)
3b. Root tubers neither spinous nor edible .... *Dioscorea deltoidea*

51. FAMILY – Dipterocarpaceae (SARJA Kul.)

1a. Calyx partly enlarged in fruit 2
1b. Calyx scarcely enlarged in fruit .... *Vateria indica* (Ashwakarna)
2a. Calyx in fruit with a distinct tube; only two calyx lobes enlarged .... *Dipterocarpus alatus* (Kashthatail)
2b. Calyx in fruit without a tube; 3-5 calyx lobes enlarged........ *Shorea robusta* (Ashwakarna, Sarja)

52. FAMILY – Ebenaceae (TINDUK Kul.)

1a. Evergreen trees; leaves oblong- lanceolate .... *Diospyros peregrina* (Tinduk)
1b. Deciduous trees; leaves ovate- lanceolate 2
2a. Leaves acute at apex .... *Diospyros melanoxylon* (Kaktinduk)
2b. Leaves obtuse at apex .... *Diospyros exculpta*

53. FAMILY – Ehretiaceae (BAHUWAR Kul.)

1a. Trees; style 4- fid; fruits with sticky juice 2
1b. Trees or shrubs; style 2 fid; fruits not with sticky juice 5
2a. Leaves alternate, broadly ovate or orbicular, 3-5 nerved from base 3
2b. Leaves subopposite, oblong to oblanceolate; one nerved .... *Cordia sinensis* (Gondani)
3a. Filaments hairy at base 4
3b. Filaments glabrous at base .... *Cordia monoica*
4a. Calyx not ribbed .... *Cordia dichotoma* (Bahuwar)
4b. Calyx ribbed .... *Cordia macleodii* (Dahiwan)
5a. Leaves 8-12 cm long, glabrous beneath .... *Ehretia laevis* (Ajan vriksha)
5b. Leaves 5-8 cm long, hairy beneath .... *Ehretia aspera*

54. FAMILY – Ephedraceae (HUM Kul.)

Ephedra gerardiana (HUM)

55. FAMILY – Euphorbiaceae (ERAND Kul.)

1a. Flowers in cyathium inflorescence 2
1b. Flowers not in cyathium inflorescence 8
2a. Shrubs with green stem and branches; leaves thick or absent 3
2b. Herbs with brownish stem and branches; leaves thin 6

3a. Plants with several branches, without spines; leaves small, early deciduous .... *Euphorbia tirucalli* (Sher)
3b. Plants spinous; leaves large, deciduous or not 4
4a. Branches 3- winged .... *Euphorbia antiquorum* (Vajri)
4b. Branches not winged 5
5a. Spines arranged in vertical lines on the stem; leaves oblanceolate .... *Euphorbia nivulia* (Sehund)
5b. Spines arranged in spiral lines on the stem; leaves lanceolate .... *Euphorbia neriifolia* (Sehund)
6a. Plants erect 7
6b. Plants prostrate .... *Euphorbia thymifolia* (Laghu Dugdhika)
7a. Plants hairy .... *Euphorbia hirta* (Nagarjuni)
7b. Plants glabrous .... *Euphorbia indica* (Dugdhika)
8a. Petals present, at least in male flowers........ 9
8b. Petals absent 17
9a. Ovary 2- locular, fruit 2-4- seeded 10
9b. Ovary 3- locular, fruits 1-6 or more seeded 12
10a. Plant juice milky; leaves 5-7- nerved from base .... *Aleurites moluccana* (Jangli Akrod)
10b. Plant juice watery; leaves 1- nerved. 11
11a. Calyx hairy; fruits oblong .... *Bridelia crenulata*
11b. Calyx glabrous; fruits globose .... *Bridelia montana* (Ekavir)
12a. Shrubs or trees 13
12b. Herbs 16
13a. Flowers in 2-3- chotomous cymes 14
13b. Flowers in spikes or racemes 15
14a. Leaves green, not glandular-hairy; flowers greenish - yellow .... *Jatropha curcas* (Mogli Erand)
14b. Leaves bronze- red, glandular hairy along margins; flowers red. .... *Jatropha gossypifolia* var. *elegans* (Lal Erand)
15a. Leaves ovate, serrate, membranous, 3-5- nerved from the base; capsules ovoid, 2-2.5 cm long .... *Croton tiglium* (Jaypal)
15b. Leaves oblong- lanceolate, crenate, thick-coriaceous, 1- nerved; capsules globose, 1-1.5 cm in diameter .... *Croton oblongifolius* (Nagdanti)
16a. Erect; leaves 2 – glandular at base; styles red .... *Chrozophora rottleri* (Patthar phod)
16b. Prostrate; leaves eglandular at base; styles yellow .... *Chrozophora prostrata* (Patthar phod)
17a. Plants with stinging hairs .... *Tragia plukenetii* (Vrishchikali)
17b. Plants without stinging hairs 18

18a. Plants with thorny branches; fruits globose; 5-6 mm in diameter; white, shining .... *Securinega leucopyros* (Kate Puvan)

18b. Plants and fruits not as above 9

19a. Trees, shrubs or under shrubs 20

19b. Herbs 28

20a. Stout-undershrubs; leaves 3-5 nerved from the base, 2- glandular at base; flowers in terminal racemes or panicles, androgynous .... *Baliospermum solaniifolium* (Danti)

20b. Plants, leaves and flowers not as above 21

21a. Plants monoecious 22

21b. Plants dioecious 26

22a. Plants with acrid, milky juice, growing along sea shores .... *Excoecaria agallocha* (Hura)

22b. Plants with watery juice, growing inland 23

23a. Stem hollow; leaves palmately lobed; stamens branched; fruits with long soft spine- like processes .... *Ricinus communis* (Erand)

23b. Stem solid; leaves, stamens and fruits not as above.. 24

24a. Trees; fruits greenish- yellow when ripe. 25

24b. Shrubs; fruits black when ripe.. *Phyllanthus reticulatus* (Krishnakamboji)

25a. Leaves 1-1.5 cm. long, obtuse; flowers in axillary clusters; ovary 3- locular; fruits globose, 2-6 cm in diameter. .... *Phyllanthus emblica* (Amalki)

25b. Leaves 5-8 cm long, acute; flowers in slender racemes on old wood; fruits depressed globose, 1.2 –2 cm in diameter, angled .... *Phyllanthus acidus* (Lavali)

26a. Leaves 3-5 nerved from the base; flowers in spikes or racemes; stamens many; fruits 2-4 seeded 27

26b. Leaves 1- nerved; flowers axillary, solitary or in clusters; stamens 2-4, fruits 1 – seeded.............................. *Drypetes roxburghii* (Putrajeev)

27a. Leaves opposite; male inflorescence pendulous, fruits 2-4 cm in diameter not covered with red powder.. *Trewia nudiflora* (Pindar)

27b. Leaves alternate or subopposite; male inflorescence erect; fruits 8-12 mm in diameter, covered with bright red powder .... *Mallotus phillippensis* (Kampillak, Kapila)

28a. Leaves distichous, entire, elliptic – oblong, 6-12 mm long, obtuse; flowers in axillary clusters; ovary locules 2–ovuled.. *Phyllanthus amarus* (Bhumyamalki)

28b. Leaves not distichous, serrulate or serrate; flowers in racemes; ovary locules 1-ovuled 29

29a. Plants hairy; leaves ovate, serrate; female flowers subtended by large bract; fruits hairy .... *Acalypha indica* (Haritmanjari)

29b. Plants glabrous; leaves linear- oblong, serrulate; female flowers not subtended by large bract; fruits spinulose on the back .... *Sebastiania chamaelea* (Bhui Erand)

56. FAMILY - Euryalaceae (MAKHANNAM Kul.)

Euryale ferox (Makhannam)

57. FAMILY – Fabaceae (APARAJITA Kul.)

1a. Trees 2

1b. Habit otherwise, not trees 12

2a. Leaves 3- foliolate.......... 3

2b. Leaves pinnate 6

3a. Petals subequal in length; pods flattened 4

3b. Petals very unequal in length; pods- turgid, moniliform.......... 5

4a. Petals white or pale pink; pods distinctly jointed .... *Desmodium oogeinensis* (Tinish)

4b. Petals orange or salmon – coloured, rarely white or pale yellow; pods not jointed .... *Butea monosperma* (Palash)

5a. Plants cultivated; stem smooth or prickly, greenish; leaves thin, more or less glabrous .... *Erythrina variegata* (Paribhadra)

5b. Plants wild; stem with cracked, brownish bark; leaves thick, densely hairy beneath .... *Erythrina suberosa* (Paribhadra)

6a. Leaflets 3-9 7

6b. Leaflets more than 9 10

7a. Petals yellow; pods flat, orbicular, winged, one seeded 8

7b. Petals white, pink or cream- coloured; pods not as above 9

8a. Stem brownish inside; leaflets 5-9 .... *Pterocarpus marsupium* (Bijak)

8b. Stem dark – red inside; leaflets 3. .... *Pterocarpus santalinus* (Raktachandan)

9a. Leaflets ovate – oblong, acute; petals white or pink; pods turgid, woody.......... *Pongamia pinnata* (Karanj)

9b. Leaflets obovate, acuminate; petals cream – coloured; pods flat; membranous .... *Dalbergia sissoo* (Shisam)

10a. Stamens monadelphous or if diadelphous, then 5 + 5; pods flat .... *Dalbergia latifolia* (Shisam)

10b. Stamens diadelphous (9 + 1), pods terete or 4 – angled 11

11a. Flowers white, 6-15 cm long; pods 4 angled .... *Sesbania grandiflora* (Agastya)
11b. Flowers yellow, 1-2 cm long; pods terete .... *Sesbania sesban* (Jayanti)
12a. Flowers yellow 13
12b. Flowers not yellow 22
13a. Leaves simple or unifoliolate .... *Crotalaria juncea* (Shan)
13b. Leaves 3- foliolate; or pinnate; 14
14a. Leaves 3- foliolate; pods aerial 15
14b. Leaves pinnate with two pairs of leaflets; pods underground .... *Arachis hypogaea* (Bhuimug)
15a. Plants prostrate, suberect or twining 16
15b. Plants erect 21
16a. Calyx persistent and much enlarged in fruit .... *Paracalyx scariosa* (Ran Ghewda)
16b. Calyx not as above 17
17a. Pods flattened .... *Macrotyloma uniflora* (Kulattha)
17b. Pods terete 18
18a. Leaflets deeply lobed 19
18b. Leaflets not or only shallowly lobed 20
19a. Wild plants; lobes of leaflets obtuse or subacute .... *Vigna trilobata* (Mugdaparni)
19b. Cultivated plants; lobes of leaflets acute .... *Vigna aconitifolia* (Makustha)
20a. Pods with long spreading hairs; seeds black. .... *Vigna mungo* (Masha)
20b. Pods with short hairs; seeds green or yellowish green .... *Vigna radiata* (Mugda)
21a. Undershrubs; pods flattened .... *Cajanus cajan* (Adhaki)
21b. Tiny herbs; pods turgid .... *Melilotus indica* (Ranmethi)
22a. Plants scandent, twining, creeping or trailing 23
22b. Plants erect or diffuse 35
23a. Leaves 3- foliolate 24
23b. Leaves pinnate 31
24a. Huge scandent shrubs with large underground tuber; flowers bluish .... *Pueraria tuberosa* (Vidari Kand, Bender)
24b. Plants not as above 25
25a. Flowers dark purple in drooping racemes; pods "S" – shaped, clothed with irritant bristles .... *Mucuna pruriens* (Kapikachhu)
25b. Flowers and pods not as above 26
26a. Pods flat or 4- gonous 27
26b. Pods terete 30
27a. Pods thick, 4- gonous, sword- shaped, hard .... *Canavalia glandiata* (Abai)
27b. Pods not as above 28

28a. Plants slender, creeping or trailing; flowers white, turning brick red; pods narrow with curved beak .... *Teramnus labialis* (Mashparni)

28b. Plants rather stout; flowers and pods not as above 29

29a. Flowers bracteolate; keel spirally twisted; stigma oblique .... *Phaseolus vulgaris* (Shrawan Ghewda)

29b. Flowers ebracteolate; keel not twisted; stigma terminal........ *Lablab purpureus* (Nishpaw, Wal)

30a. Pods 20-30 cm long, pale green .... *Vigna sinensis* (Rajmash)

30b. Pods 8-15 cm long, dark green .... *Vigna cylindrica*

31a. Terminal leaflets modified into tendrils 32

31b. Terminal leaflets not so modified 34

32a. Wings free from the keel .... *Lathyrus sativus* (Lanka)

32b. Wings adherent to the keel 33

33a. Flowers mostly white; seeds globose, uniformly green........ *Pisum sativum* (Matar)

33b. Flowers white or coloured; seeds rather compressed, grey mottled with green .... *Pisum arvense* (Satin, Watana)

34a. Leaflets 5-7, alternate; flowers solitary, axillary, blue or white; seeds brown .... *Clitoria ternatea* (Gokarna, Aparajita)

34b. Leaflets 20-30, opposite; flowers in axillary racemes, pink or white; seeds red or white with a black spot .... *Abrus precatorius* (Gunja)

35a. Leaves 1 to 3-foliolate 36

35b. Leaves pinnate 42

36a. Thorny undershrubs; pods torulose .... *Alhagi pseudalhagi* (Yawas)

36b. Plants not thorny; pods not torulose 37

37a. All leaves one – foliolate 38

37b. All leaves 3- foliolate or 1 and 3 foliolate leaves intermixed 40

38a. Pods not jointed, turgid 39

38b. Pods jointed, flattened .... *Desmodium gangeticum* (Shalparni)

39a. Pods 1- seeded, gland- dotted .... *Psoralea coryllifolia* (Bakuchi)

39b. Pods many- seeded, not gland – dotted .... *Crotalaria verrucosa* (Shanpushpi)

40a. All leaves 3- foliolate; pods not jointed, 4-gonous 41

40b. One and 3 – foliolate leaves intermixed; pods jointed; joints folded .... *Uraria lagopoides* (Prishniparni)

41a. Flowers in dense axillary racemes .... *Cyamopsis tatragonoloba* (Gawar)

41b. Flowers axillary, solitary........ *Trigonella foenum- greaceum* (Methika)

42a. Leaflets 3-9; pods jointed; joints folded .... *Uraria picta* (Prishniparni)

42b. Leaflets more than 9; pods not as above 43

43a. Leaflets serrate- margined; flowers axillary,
solitary or 2-3 together, pods 1-2-seeded ....*Cicer arietinum* (Chanak)

43b. Leaflets entire; flowers in axillary
racemes; pods 3-many seeded 44

44a. Leaves gland – dotted; pods echinate,
3-4 seeded ....*Glycirrhiza glabra* (Yashtimadhu)

44b. Leaves not gland – dotted; pods not echinate,
4-many-seeded 45

45a. Lateral veins of leaflets parallel,
V- shaped; pods flat ....*Tephrosia purpurea* (Sharpunkha)

45b. Lateral veins of leaflets not parallel, pods cylindric ... .*Indigofera tinctoria* (Nili)

58. FAMILY - Fagaceae (MAYAPHAL Kul.)

Quercus infectoria (Mayaphal)

59. FAMILY – Flacourtiaceae (PRACHINAMALAK Kul.)

1a. Plants thorny; petals absent ....*Flacourtia indica* (Prachinamalak)

1b. Plants not thorny; petals 5-..*Hydnocarpus laurifolius* (Katukapittha)

60. FAMILY – Fumariaceae (PARPAT Kul.)

Fumaria indica (Parpat)

61. FAMILY – Gentianaceae (CHIRAIT Kul.)

1a. Flowers in axillary clusters ....*Enicostema axillare* (Mamijwa)

1b. Flowers in terminal cymes 2

2a. Petals 4, unequal, without pits or glands at base ....*Canscora decussata* (Shankhahuli)

2b. Petals 4-5, equal with glands at base 3

3a. Petals yellow; a high altitude plant ....*Swertia chirayita* (Chirattikta)

3b. Petals white; a low altitude plant ....*Swertia densiflora*

62. FAMILY - Hippocrateaceae (SAPTAACHAKRA Kul.)

Salacia chinensis (Saptachakra)

63. FAMILY – Hypoxidaceae (TALMULI Kul.)

Curculigo orchioides (Talmuli)

64. FAMILY – Iridaceae (KESHAR Kul.)

1a. Flowers solitary, sessile on the underground corm; ....*Crocus sativa* (Keshar)

1b. Flowers solitary or in pairs, stalked on a
scape; corolla tube absent or very short ....*Iris germanica* (Hemvati, Pushkarmul)

65. FAMILY – Juglandaceae (AKROT Kul.)

Juglans regia (Akrot)

66. FAMILY – Lamiaceae (TULSI Kul.)

1a. Flowers actinomorphic with 4 equal petals, in axillary clusters or terminal spikes 2
1b. Flowers zygomorphic, 5- merous, in terminal thyrsoid panicles or verticillasters 3
2a. Flowers all in axillary whorls, pedicelled .... *Mentha arvensis* (Putinash karpur)
2b. Flowers in axillary as well as in terminal whorls, sessile .... *Mentha spicata* (Pudina)
3a. Calyx 2 - lipped 4
3b. Calyx tubular 9
4a. Flowers pedicelled 5
4b. Flowers sessile in thick clustered spikes .... *Majorana hortensis* (Marva)
5a. Upper lip of calyx with large deflexed single lobe, lower lip 4-lobed; lower lip of corolla 1 – lobed, more or less flat 6
5b. Upper lip of calyx with 2 narrow lobes, lower lip 3- lobed; lower lip of corolla 3- lobed, mid- lobe boat – shaped 8
6a. Pedicels longer than the calyx; corolla pink .... *Ocimum tenuiflorum* (Tulsi)
6b. Pedicels shorter than the calyx; corolla not pink 7
7a. Corolla white .... *Ocimum basilicum* (Barbar)
7b. Corolla yellow .... *Ocimum gratissimum* (Ramtulsi)
8a. Roots not tuberous; leaves strongly aromatic; corolla blue .... *Coleus amboinicus* (Panacha Onwa)
8b. Roots tuberous; leaves not strongly aromatic; corolla white or pink .... *Coleus barbatus* (Mainmul)
9a. Calyx 5-toothed; corolla 4- lobed, lobes unequal; stamens not didynamous, equal or subequal .... *Pogostemon parviflorus* (Phanijjak)
9b. Calyx 10-15 toothed; corolla 2- lipped and 5-lobed, stamens didynamous 10
10a. Upper lip of corolla short, nearly flat .... *Anisomeles malabarica* (Sprukka)
10b. Upper lip of corolla hooded 11
11a. Flowers in secund terminal spikes; corolla blue .... *Hyssopus officinalis* (Jufa)
11b. Flowers in axillary or terminal verticils; corolla not blue 12
12a. Verticils large, many, spinous; corolla velvety – orange .... *Leonotis napetifolia* (Matisul)
12b. Verticil rather small, solitary, not spinous; corolla white .... *Leucas cephalotes* (Droanpushpi)

67. FAMILY – Lauraceae (KARPUR Kul.)

1a. Leaves stiff, shining, 3-5- nerved from the base 2

1b. Leaves not as above, 1- nerved .... *Litsaea glutinosa*

2a. Leaves broad 3

2b. Leaves narrow .... *Cinnamomum tamala* (Tamal Patra)

3a. Flower buds enclosed by large imbricated scales.................. *Cinnamomum camphora* (Karpur)

3b. Flower buds naked or with very small scales .... *Cinnamomum verum* (Twak)

68. FAMILY - Lecythidaceae (SAMUDRA PHAL Kul.)

1a. Leaves acute; racemes pendulous 2

1b. Leaves obtuse; racemes erect .... *Careya arborea* (Kumbhi)

2a. Leaves up to 30 cm long; fruit ovoid ....*Barringtonia racemosa* (Samudra Phal)

2b. Leaves upto 15 cm long, fruits 4- angled ....*Barringtonia acutangula* (Hijjal)

69. FAMILY – Liliaceae (VANPALANDU Kul.)

1a. Plants scandent or climbing 2

1b. Plants erect 3

2a. Leaves reduced to spines; branchlets green,
leaf like (cladode); flowers small, white ....*Asparagus racemosus* (Shatawari)

2b. Leaves well- developed with spring-like
coiled tips; flowers large, yellow-red ....*Gloriosa suberba* (Kalikari)

3a. Plants as in 2 a ....*Asparagus adscendens* (Shwet Musali)

3b. Plants not as above 4

4a. Leaves thick, fleshy, with spinulose margins;
flowers yellowish – red on a tall branched scape ....*Aloe vera* (Kumari)

4b. Leaves and flowers not as above 5

5a. Plants with underground bulb or corm 6

5b. Plants with thick tuberous roots ....*Chlorophytum arundinaceum*
(Saphed Musali)

6a. Plants with underground corms; flowers yellow.................. *Colchicum luteum* (Hiranyatuttha)

6b. Plants with underground bulbs; flowers
pink or brownish- yellow 7

7a. Leaves oblanceolate, green with brown spots;
flowers pink ....*Scilla hyacinthina* (Ran Kanda)

7b. Leaves linear-oblong, green without brown
spots; flowers brownish-yellow ....*Drimia indica* (Vanpalandu)

70. FAMILY - Linaceae (ATASI Kul.)

Linum usitatissimum (Atasi)

71. FAMILY - Lobeliaceae (NAL Kul.)

Lobelia nicotianaefolia (Nal, Devnal)

72. FAMILY - Loranthaceae (BANDAK Kul.)

1a. Plants with leaves; flowers greenish or red with long perianth tube .... *Dendrophthoe falcata* (Bandak)

1b. Plants without leaves; flowers greenish, without perianth tube .... *Viscum nepalense* (Hadjodi)

73. FAMILY - Lythraceae (MENDIKA Kul.)

1a. Herbs; flowers minute, red in axillary clusters .... *Ammannia baccifera* (Agnipatri)

1b. Trees or shrubs; flowers not as above 2

2a. Trees with ash- green leaves and white flowers .... *Lagerstroemia parviflora* (Lendi)

2b. Shrubs with green leaves; flowers cream- colored or red 3

3a. Wild plants; leaves bifarious; flowers red in axillary cymes .... *Woodfordia fruticosa* (Dhataki)

3b. Cultivated plants; leaves not bifarious; flowers cream-coloured .... *Lawsonia inermis* (Mendika)

74. FAMILY – Magnoliaceae (CHAMPAK Kul.)

Michelia champaca (Suwarn Champak)

75. FAMILY - Malvaceae (KARPAS Kul.)

1a. Trees with white, polygamous flowers in panicles; involucral bracts persistent in fruit .... *Kydia calycina*

1b. Shrubs, undershrubs or herbs; flowers bisexual, axillary solitary or few; involucral bracts persistent or not in fruit 2

2a. Carpels more or less free; fruits separating into carpels 3

2b. Carpels united, fruits capsular 7

3a. Carpels 15 or more 4

3b. Carpels less then 15 5

4a. Plants cultivated; petals bluish or rose coloured .... *Althaea officinalis* (Khatmi)

4b. Plants wild; petals yellow .... *Abutilon indicum* (Atibala)

5a. Stipules spinous; leaves ovate-ellipsoid, glaucous green; corolla white .... *Sida spinosa* (Kharyashtika)

5b. Stipules not spinous; leaves ovate-cordate or lanceolate, green; corolla yellow 6

6a. Leaves ovate- cordate .... *Sida cordifolia* (Bala)

6b. Leaves lanceolate .... *Sida acuta* (Nagbala)

7a. Bracteoles 3, broad, cordate; ovary 3-5- locular; capsules ovoid; seeds with long cottony fibres .... *Gossypium arboreum* (Karpas)

7b. Bracteoles 4 or more, linear or lanceolate; ovary 5-6 locular; capsules ovoid or elongate; seeds not with long fibres 8

8a. Calyx spathaceous, split down on one side 9

8b. Calyx tubular 10

9a. Capsule 10-30 cm long, 5- angular; seeds not scented when bruised .... *Abelmoschus esculentus* (Tindish)

9b. Capsules 5-8 cm long; seeds scented when bruised .... *Abelmoschus moschatus* (Latakasturi)

10a. Stems prickly; upper leaves deeply palmately lobed; flowers yellow with brown centre .... *Hibiscus cannabinus* (Ambastha)

10b. Stems not prickly; all leaves ovate, shallowly lobed or not; flowers not as above 11

11a. Leaves shining green, smooth; flowers red pink of white .... *Hibiscus rosa-sinensis* (Japa)

11b. Leaves dull green, rough, flowers bluish or white with deep red centre .... *Hibiscus syriacus* (Japa)

76. FAMILY – Meliaceae (NIMB Kul.)

1a. Leaflets entire 2

1b. Leaflets serrate 3

2a. Anthers 5; staminal tube long .... *Aglaia elaeagnoidea* (Priyangu)

2b. Anthers 10; staminal tube short .... *Soymida febrifuga* (Rohini)

3a. Leaves unipinnate; flowers white .... *Azadirachta indica* (Nimb)

3b. Leaves 2-3 pinnate; flowers pink- purple .... *Melia azedarach* (Mahanimb)

77. FAMILY - Menispermaceae (GUDUCHI Kul.)

1a. Ovaries 3 or more 2

1b. Ovary solitary 3

2a. Leaves ovate, not cordate; seeds horse-shoe-shaped .... *Cocculus hirsutus* (Patal Garudi)

2b. Leaves broadly ovate, cordate; seeds globose............................... *Tinospora cordifolia* (Guduchi)

3a. Sepals free; petals connate .... *Cissampelos pareira* (Laghupatha)

3b. Sepals connate; petals free .... *Cyclea peltata* (Rajpatha)

78. FAMILY - Mimosaceae (BABBUL Kul.)

1a. Spreading undershrubs; leaves sensitive to touch; flowers pink in globose heads; pods 2-3 cm long .... *Mimosa pudica* (Lajjalu)

1b. Erect or scandent shrubs or trees; leaves not sensitive to touch; flowers pink, white or yellow, in globose heads or elongate spikes; pods more than 5 cm long 2

2a. Huge scandent shrub; flowers white; pods upto 1 m long .... *Entada pursaetha* (Garbeej)

2b. Plants and pod not as above 3

3a. Plants armed with prickles or spines 4

3b. Plants not armed 9

4a. Flowers in elongate spikes 5

4b. Flowers in globose heads .... *Acacia nilotica* (Babbul)

5a. Flowers of two different colours, pink and yellow in the same spike .... *Dichrostachys cinerea* (Sigamkati)

5b. Flowers of one colour in the spike 6

6a. Stamens 8-10; anthers gland-crested .... *Prosopis cineraria* (Shami)

6b. Stamens many; anthers not gland-crested 7

7a. Pinnae 20-40 pairs 8

7b. Pinnae 3-6 pairs .... *Acacia senegal* (Gond)

8a. Leaf- rachis, calyx and corolla pubescent .... *Acacia catechu* (Khadir)

8b. Leaf- rachis, calyx and corolla glabrous .... *Acacia chundra* (Lal khair)

9a. Flowers greenish- white .... *Albizia lebbeck* (Shirish)

9b. Flowers pink .... *Albizia odoratissima* (Kala Shirish)

79. FAMILY – Moraceae (WAT Kul.)

1a. Flowers in hypanthodium inflorescence 2

1b. Flowers in catkins or spikes 10

2a. Tree with long, stout aerial roots; fruits orange-red ... . *Ficus benghalensis* (Wat)

2b. Trees or shrubs with, slender aerial roots or aerial roots absent; fruits not orange-red 3

3a. Leaves caudate- acuminate at apex 4

3b. Leaves not caudate acuminate at apex 5

4a. Leaf- acumen 1-3 cm long; fruits globose, nearly black ... .*Ficus rumphii*

4b. Leaf-acumen 3-6 cm or more long; fruits depressed globose, purple ....*Ficus religiosa* (Ashwattha)

5a. Fruit sessile 6

5b. Fruits stalked 8

6a. Leaves elliptic, ovate or obovate, 5-10 cm long; fruits yellow or reddish when ripe; aerial roots few ....*Ficus microcarpa* (Nandivruksha)

6b. Leaves ovate, 7-15 cm long; fruits purplish or reddish; aerial roots absent 7

7a. Nerves of leaves indistinct; male sepals 3 ....*Ficus amplissima* (Plaksha)

7b. Nerves of leaves distinct; male sepals 4 or 5 ....*Ficus virens* (Plaksha)

8a. Leaves opposite; fruits globose, green axillary or on branches near the root . ...*Ficus hispida* (Kakodumbar)

8b. Leaves alternate; fruits pyriform, red or purplish, axillary or on branches on old stems 9

9a. Small trees, leaves broadly ovate, 3-5 lobed, serrate; fruits purplish- green, axillary .... *Ficus carica* (Anjeer)

9b. Large trees; leaves elliptic, entire, fruits brick-red on branches from old stems ....*Ficus racemosa* (Udumbar)

10a. Large trees; leaves entire; fruits large echinate, on old stems ....*Artocarpus heterophyllus* (Panas)

10b. Small trees, leaves serrate; fruits 1-3 cm long, fleshy, black on slender branches ....*Morus alba* (Tut)

80. FAMILY – Moringaceae (SHIGRU Kul.)

Moringa oleifera (Shigru)

81. FAMILY – Musaceae (KADALI Kul.)

Musa paradisiaca (Kadali)

82. FAMILY – Myricaceae (KATUPHAL Kul.)

Myrica esculenta (Katuphal)

83. FAMILY – Mýristicaceae (JATIPHAL Kul.)

Myristica fragrans (Jatiphal)

84. FAMILY – Myrsinaceae (WIDANG Kul.)

Embelia ribes (Widang)

85. FAMILY – Myrtaceae (LAWANG Kul.)

1a. Leaves alternate, pendulous; fruits capsular 2

1b. Leaves opposite, spreading; fruits fleshy 3

2a. Bark bluish- white; flowers 1-3 together; operculum single .... *Eucalyptus globulus* (Nilgiri)

2b. Bark reddish-white; flowers in panicles; operculum double .... *Eucalyptus citriodora* (Nilgiri)

3a. Flowers large, 1-3 together; ovary 4-5 locular; fruits greenish-yellow .... *Psidium guayava* (Peruk)

3b. Flowers small in cymes; ovary 2-locular; fruits dark pruple 4

4a. Petals united into a calyptra .... *Syzygium cumini* (Jambu)

4b. Petals distinct .... *Syzygium aromaticum* (Lawang)

86. FAMILY – Naucleaceae (KADAMB Kul.)

1a. Heads 2-2.5 cm in diam., corolla lobes valvate in bud .... *Haldina cordifolia* (Haldu)

1b. Heads 2.5-4 cm in diam.; corolla lobes imbricate in bud .... *Neolamarckia chinensis* (Kadamb)

87. FAMILY – Nelumbonaceae (PUNDARIK Kul.)

Nelumbo nucifera (Padma, Kamal)

88. FAMILY – Nyctaginaceae (PUNARNAVA Kul.)

1a. Spreading or trailing herbs; leaves unequal in pairs; flowers small in umbels, not surrounded by calyx- like involucre 2

1b. Erect herbs; leaves equal in pairs; flowers large, solitary, surrounded by calyx-like involucre .... *Mirabilis jalapa* (Gulbakshi)

2a. Leaves white beneath; flowers sessile .... *Boerhavia diffusa* (Punarnava)

2b. Leaves not white beneath; flowers pedicelled .... *Boerhavia chinensis* (Mothi Punarnava)

89. FAMILY – Nymphaeaceae (KAMAL Kul.)

1a. Leaves with entire margin 2

1b. Leaves with dentate or sinuate margin 3

2a. Leaves pubescent beneath; flowers opening in the evening; petals pink or red .... *Nymphaea pubescens* (Raktakamal)

2b. Leaves glabrous beneath; flowers opening in the morning; petals white .... *Nymphaea alba* (Shwetkamal)

3a. Flowers opening in the morning; petals white or bluish .... *Nymphaea nouchali* (Shwetkamal)

3b. Flowers opening in the evening; petals red .... *Nymphaea rubra* (Raktakamal)

90. FAMILY – Oleaceae (PARIJATAK Kul.)

1a. Leaves simple 2
1b. Leaves compound 4
2a. Trees; corolla tube orange coloured .... *Nyctanthes arbor-tristis* (Parijatak)
2b. Erect or scandent shrubs; corolla tube not orange coloured 3
3a. Flowers normal; many together .... *Jasminum multiflorum* (Kund)
3b. Flowers double, 3-together .... *Jasminum sambac* (Mallike)
4a. Trees; fruits capsular .... *Schrebera swietenioides*
4b. Scandent or twining shrubs; fruits fleshy 5
5a. Leaves 3-foliolate 6
5b. Leaves pinnate 7
6a. Side leaflets much smaller than the terminal one or absent .... *Jasminum auriculatum* (Jatika)
6b. Side leaflets slightly smaller or equal to the terminal one .... *Jasminum calophyllum*
7a. Flowers white without purple tinge .... *Jasminum officinale* (Chameli)
7b. Flowers white with purple stripes .... *Jasminum grandiflorum* (Jati)

91. FAMILY – Orchidaceae (MALAKAND Kul.)

1a. Terrestrial orchids with underground tubers 2
1b. Epiphytic, without tubers .... *Vanda tessellata*
2a. Underground tubers palmately lobed; flowers white .... *Orchis latifolia* (Beejgandh)
2b. Underground tubers unlobed; flowers purplish-yellow .... *Eulophia nuda* (Malakand)

92. FAMILY – Oxalidaceae (CHANGERI Kul.)

1a. Leaves 3-foliolate; flowers in lax racemes .... *Oxalis corniculata* (Changeri)
1b. Leaves pinnate; flowers in terminal clusters .... *Biophytum sensitivum* (Lajari)

93. FAMILY – Pandanaceae (KETAKI Kul.)

Pandanus odoratissimus (Ketaki) (=*P. fascicularis*)

94. FAMILY – Papaveraceae (AHIFEN Kul.)

1a. Plant juice yellow; leaves and capsules spinous; flowers yellow .... *Argemone mexicana* (Satyanashi, Swarnakshiri)

1b. Plant juice milky; leaves and capsules not spinous; flowers red or pink .... *Papaver somniferum* (Ahifen)

95. FAMILY – Parmeliaceae (SHAILEYA Kul.)

Parmelia perlata (Shaileya)

96. FAMILY – Pedaliaceae (Til Kul.)

1a. Flowers yellow; fruits drupaceous, 4-spinous .... *Pedalium murex* (Gokshur)

1b. Flowers white or pink; fruits capsular, not spinous .... *Sesamum orientale* (Til)

97. FAMILY – Periplocaceae (SARIWA Kul.)

1a. Flowers dull- brown outside, red within; corolla lobes valvate .... *Hemidesmus indicus* (Krishna Sariwa, Uplasari)

1b. Flowers greenish- white, corolla lobes twisted and overlapping .... *Cryptolepis buchanani* (Shwet sariva)

98. FAMILY – Pinaceae (DEODARU Kul.)

1a. Leaves 3 in a cluster with basal sheaths .... *Pinus roxburghii* (Saral)

1b. Leaves 10 or more in a cluster, without basal sheaths .... *Cedrus deodora* (Deodaru)

99. FAMILY – Piperaceae (MARICH Kul.)

1a. Leaves cordate at base 2

1b. Leaves acute or rounded at base 3

2a. Leaves broadly ovate; fruits black .... *Piper longum* (Pippali)

2b. Leaves elliptic – ovate; fruits brown .... *Piper cubeba* (Kankol)

3a. Leaves rounded at base, broadly ovate, fruits red, turning black .... *Piper nigrum* (Marich)

3b. Leaves acute at base; ovate- oblong; fruits greenish black .... *Piper betle* (Nagvalli)

100. FAMILY – Pistaciaceae (RUMKUNDRU Kul.)

Pistacia vera (Rumkundru)

101. FAMILY – Plantaginaceae (ISHATGOL Kul.)

Plantago ovata (Ishatgol)

102. FAMILY – Plumbaginaceae (CHITRAK Kul.)

1a. Flowers white .... *Plumbago zeylanica* (Chitrak)

1b. Flowers not white 2

2a. Flowers blue .... *Plumbago capensis* (Kala Chitrak)

2b. Flowers red or pink .... *Plumbago indica* (Lal Chitrak)

103. FAMILY – Poaceae (TRUNDHANYA Kul.)

1a. Plants or their parts aromatic 2
1b. Plants or their parts not aromatic 5
2a. Roots aromatic .... ***Vetiveria zizanioides*** **(Ushir)**
2b. Leaves and inflorescence aromatic 3
3a. Leaves smelling of lemon .... ***Cymbopogon citratus*** **(Sugandh Bhutruna)**
3b. Leaves not with the above smell 4
4a. Leaves smelling of odomos .... ***Cymbopogon winterianus*** **(Cetronella)**
4b. Leaves smelling of lemon and rose mixed; a wild grass .... ***Cymbopogon martinii*** **(Rohish)**
5a. Short perennial grass with running stem; inflorescence of 3-5 digitately spreading tiny racemes .... ***Cynodon dactylon*** **(Durva)**
5b. Tall, robust grasses; stem not running; inflorescence not as above 6
6a. Inflorescence large, feather- shaped, silky hairy 7
6b. Inflorescence narrow, columnar, short hairy .... ***Desmostachya bipinnata*** **(Darbh)**
7a. Stem hollow; leaves amplexicaule at base; spikelets hairy, brownish .... ***Arundo donax*** **(Devnal)**
7b. Stem solid; leaves not amplexicaule at base; spikelets shining white or silvery 8
8a. Inflorescence silvery white 9
8b. Inflorescence purplish- brown .... ***Saccharum arundinaceum***
9a. Cultivated grass; stem with sweet juice .... ***Saccharum officinarum*** **(Ikshu)**
9b. Wild grasses; stem without sweet juice 10
10a. Two or four m. tall grasses .... ***Saccharum spontaneum*** **(Shar)**
10b. Upto 1.5 cm tall grasses .... ***Saccharum munja*** **(Shar)**

104. FAMILY – Polygonaceae (CHUKRA Kul.)

1a. Stout herb with large broad basal leaves; stamens usually 9 .... ***Rheum emodi*** **(Peetmuli)**
1b. Small herb with many stem leaves; stamens usually 6 .... ***Rumex vesicarius*** **(Chukra)**

105. FAMILY – Portulacaceae (LONIKA Kul.)

1a. Nodes glabrous; leaves 5-50 mm long, alternate; flowers clustered .... ***Portulaca oleracea*** **(Bruhlloni)**
1b. Nodes hairy, leaves 3-10 mm long, opposite; flowers solitary .... ***Portulaca quadrifida*** **(Loni)**

106. FAMILY – Ranunculaceae (WATSANABH Kul.)

1a. Climbing shrubs or undershrubs; flowers apetalous … .... *Clematis heynei* (Moorva)
1b. Erect herbs; flowers with petals 2
2a. Roots normal; one sepal spurred;
petals 4; follicles 3 .... *Delphinium denudatum* (Jadwaar, Nirwishi)
2b. Roots thickened, one sepal hooded; petals 2;
follicles 5 3
3a. Racemes unbranched or few branches
at base; flowers pale blue .... *Aconitum ferox* (Halahal)
3b. Racemes much branched; flowers bright blue .... *Aconitum heterophyllum* (Watsanabh)

107. FAMILY – Rhamnaceae (BADAR Kul.)

1a. Leaves glabrous beneath .... *Ziziphus jujuba* (Badar)
1b. Leaves densely white hairy beneath .... *Ziziphus mauritiana* (Badar)

108. FAMILY – Rosaceae (GULAB Kul.)

1a. Gynoecium monocarpellary, superior 2
1b. Gynoecium many carpellary; carpels
enclosed in a hypanthium .... *Rosa centifolia* (Gulab)
2a. Leaves acute; flowers peduncled in
cymose umbels; fruits with juicy flesh .... *Prunus cerasioides* (Padmakashtha)
2b. Leaves acuminate; flowers sessile,
solitary; fruits with fibrous, hard flesh .... *Prunus amygdalus* (Vatam)

109. FAMILY – Rubiaceae (MANJISHTHA Kul.)

1a. Shrubs or trees; leaves stipulate, 2 or 3 at each node 2
1b. Climbing herbs; leaves exstipulate, 4-8 at each node .... *Rubia cordifolia* (Manjishtha)
2a. Plants thorny .... *Catunaregam spinosa* (Madan)
2b. Plants not thorny 3
3a. Plants resinous; ovary 1- locular with 4-5 parietal
placentas .... *Gardenia gummifera* (Nadihingu)
3b. Plants and ovary not as above 4
4a. Slender twining shrubs .... *Paederia fodtida* (Prasarani)
4b. Erect shrubs 5
5a. Leaves glabrous, mostly opposite;
flowers red in dense corymbose cymes .... *Ixora coccinea* (Pankul)
5b. Leaves hairy, mostly 3 in a whorl; flowers
white, pink or bluish in danse trichotomous
cymes .... *Spermadictyon suaveolens* (Jitsaya)

110. FAMILY – Rutaceae (SATAP Kul.)

1a. Ovary deeply 2-5 lobed; style basilar; fruits capsular 2
1b. Ovary entire; style terminal; fruits pulpy or indehiscent 3
2a. Herbs or undershrubs; flowers bisexual; ovary locules 3 or more ovuled .... *Ruta chalepensis* (Satap)
2b. Shrubs or trees; flowers polygamous; ovary locules 2 ovuled .... *Zanthoxylum armatum* (Tirphal)
3a. Leaves one foliolate; fruit juicy with leathery epicarp 4
3b. Leaves 3- foliolate; fruit pulpy with hard epicarp .... *Aegle marmelos* (Bilva)
4a. Petiole wingless, not articulated with the blade; fruits with rough thick skin 5
4b. Petiole winged, articulated with the blade; fruits with smooth thin skin .... *Citrus aurantifolia* (Nimbuk)
5a. Fruits ellipsoid .... *Citrus medica* (Matullung)
5b. Fruits globose .... *Citrus limon* (Dantshath)

111. FAMILY – Salvadoraceae (PILU Kul.)

Salvadora persica (Pilu)

112. FAMILY - Samydaceae

1a. Branchlets and leaves beneath glabrous .... *Casearia esculenta*
1b. Branchlets and leaves beneath hairy .... *Casearia elliptica* (Massi)

113. FAMILY – Santalaceae (CHANDAN Kul.)

Santalum album (Chandan)

114. FAMILY – Sapindaceae (ARISHTAK Kul.)

1a. Herbaceous climber; leaves biternate; flowers zygomorphic; fruits bladdery .... *Cardiospermum helicacabum* (Karnasphota)
1b. Trees; leaves even pinnate; flowers actinomorphic; fruits not bladdery 2
2a. Petals absent; fruit not lobed .... *Schleichera oleosa* (Koshagra)
2b. Petals present; fruit 2-3 lobed 3
3a. Leaflets acuminate, glabrous or thinly pubescent beneath .... *Sapindus laurifolius* (Arishtak)
3b. Leaflets emarginate, densely pubescent beneath, especially when young .... *Sapindus emarginatus* (Arishtak)

115. FAMILY – Sapotaceae (MADHUK Kul.)

1a. Calyx lobes 4-5; staminodes absent 2
1b. Calyx lobes 6-8; staminodes present 3
2a. Leaves 8-12 cm broad .... *Madhuca indica* (Madhuk)
2b. Leaves upto 5cm broad .... *Madhuca longifolia* (Madhuk)
3a. Petals 18-24; fruits upto 1.5 cm. in diameter, yellow or reddish yellow 4
3b. Petals 6; fruits 4-8 cm in diameter; rusty brown .... *Manilkara zapota* (Chikkoo)
4a. Flowers normally tetramerous .... *Mimusops elengi* (Bakul)
4b. Flowers normally trimerous .... *Manilkara hexandra* (Kshiri)

116. FAMILY – Saxifragaceae (PASHANBHED Kul.)

Bergenia ligulata (Pashanbhed)

117. FAMILY – Scrophulariaceae (BRAHMI Kul.)

1a. Plants growing in or near water 2
1b. Plants growing on land.................... 3
2a. Plants spreading on mud; leaves entire, oblanceolate .... *Bacopa monnieri* (Brahmi)
2b. Plants erect; leaves above water surface entire, those below dissected into linear segments .... *Limnophila indica* (Ambuj)
3a. Flowers yellow; corolla tube very short or absent; petals subequal .... *Verbascum chinense* (Kulahal)
3b. Flowers purple; corolla tube long, widening upwards; petals unequal .... *Digitalis purpurea* (Tilpushpi)

118. FAMILY – Simaroubaceae (MAHARUKH Kul.)

Ailanthus exselsa (Maharukh)

119. FAMILY – Smilacaceae (HIRANYASHAK Kul)

1a. Leaves elliptic – lanceolate, 8-12 cm. wide .... *Smilax zeylanica* (Hiranyashak)
1b. Leaves narrow – lanceolate, upto 6 cm. wide .... *Smilax lanceifolia* (Hariya Shukchin)

120. FAMILY – Solanaceae (KATANKARI Kul)

1a. Fruit a berry 2
1b. Fruit capsular 6
2a. Anthers connivent around the style, opening by pore or slit near the apex 3
2b. Anthers not connivent around the style, opening by longitudinal slit 5

3a. Stem and leaves prickly or spiny; flowers blue or purple 4

3b. Stem and leaves not prickly or spiny; flowers white........ *Solanum nigrum* (Kakmachi)

4a. Plants erect; fruits 1-1.5 cm in diameter .... *Solanum anguivi* (Bruhati)

4b. Plants spreading or procumbent; fruits 2-2.5 cm in diameter .... *Solanum virginianum* (Kantakari)

5a. Plants thinly pubescent or glabrous; flowers blue - purple or dull red, 2-2.5 cm long; fruting calyx spreading, not enclosing the fruit........ *Atropa bellandona* (Krishnaphala)

5b. Plants dense hairy; flowers pale yellow, upto 1 cm long; fruting calyx enclosing the fruit .... *Withania somnifera* (Ashwagandha)

6a. Capsules opening by lid .... *Hyoscyamus niger* (Parasik yawani)

6b. Capsules splitting lengthwise or bursting irregularly 7

7a. Capsules smooth, partly enclosed by the calyx .... *Nicotiana tabacum* (Tamra koot)

7b. Capsules prickly, not enclosed by the calyx .... *Datura metel* (Krishna Dhattur)

121. FAMILY – Sterculiaceae (MUCHKUND Kul.)

1a. Shrubs 2

1b. Trees 3

2a. Flowers on leaf-opposed peduncles, actinomorphic, purplish; stamens 3-5; capsule 5- winged .... *Abroma augusta* (Pishach Karpas)

2b. Flowers on axillary peduncles; zygomorphic, red; stamens 10; fruit formed of 5, spirally twisted follicles........ *Helicteres isora* (Avartani)

3a. Flowers unisexual, apetalous, in glandular hairy panicles; fruits of 4-6 follicles 4

3b. Flowers bisexual, with 5 petals, in axillary cymes or solitary; fruits capsular 5

4a. Leaves digitate with 5-9 leaflets; follicles smooth, not bristly .... *Sterculia foetida* (Pun)

4b. Leaves simple, palmately 5-7 – lobed; follicles densely bristly .... *Sterculia urens* (Kandol)

5a. Leaves deeply cordate and subpeltate .... *Pterosperum acerifolium* (Muchkund)

5b. Leaves rounded at base and not subpeltate .... *Plerospermum suberifolium* (Kaniar)

122. FAMILY – Strychnaceae (KARASKAR Kul.)

1a. Fruits orange, 3-8 cm in diameter, many seeded .... *Strychnos nux-vomica* (Karaskar, Wishtinduk)

1b. Fruits black, 2 cm in diameter, 1-2 seeded .... *Strychnos petatorum* (Katak, Nirmali)

123. FAMILY – Symplocaceae (LODHRA Kul.)

Symplocos racemosa (Lodhra)

124. FAMILY – Taxaceae (TALIS Kul.)

Taxus baccata (Talis)

125. FAMILY – Thymelaeaceae (AGURU Kul.)

Aquillaria agallocha (Aguru)

126. FAMILY – Tiliaceae (PURUSHAK Kul.)

1a. Trees or shrubs 2

1b. Herbs or undershrubs 7

2a. Trees; petals sepaloid; anthers opening by terminal pore; fruit fleshy with hard tuberculate stone .... *Elaeocarpus sphaericus* (Rudraksha)

2b. Shrubs or small trees 3

3a. Leaves white hairy beneath 4

3b. Leaves not white hairy beneath 5

4a. Stipules leafy, auricled .... *Grewia tiliifolia* (Dhanwan, Dhaman)

4b. Stipules linear- lanceolate .... *Grewia asiatica* (Parushak)

5a. Leaves orbicular; drupes with crustaceous rind .... *Grewia villosa* (Kharmati)

5b. Leaves not orbicular; drupes with leathery rind 6

6a. Tall shrubs; leaves broadly ovate; flowers yellow, bisexual .... *Grewia scabrophylla* (Safed Dhaman)

6b. Dwarf shrubs; leaves linear- lanceolate; flowers white (male. or yellow (bisexual), polygamous .... *Grewia polygama* (Kukur Bich)

7a. Capsules depressed globose .... *Corchorus capsularis* (Kalshak)

7b. Capsules elongate 8

8a. Capsules 3-8 cm long 9

8b. Capsules 1-1.5 cm long .... *Corchorus fascicularis* (Harankhuri)

9a. Capsules glabrous, cylindric, 10- ribbed, usually 5-valved .... *Corchorus olitorius* (Brihatchanchu)

9b. Capsules hairy, 3-angled, usually 3- valved .... *Corchorus trilocularis* (Katuchanchu)

127. FAMILY – Trapaceae (SHRUNGATAK Kul.)

Trapa natans var. *bispinosa* (Shrungatak)

128. FAMILY- Ulmaceae (CHIRBILWA Kul.)

Heloptelea integrifolia (Chirbilwa)

129. FAMILY – Valerianaceae (JATAMANSI Kul.)

Nardostachys grandiflora (Jatamansi)

130. FAMILY – Verbenaceae (NIRGUNDI Kul.)

1a. Creeping herbs; flowers in dense heads, white or purple .... *Phyla nodiflora* (Ratvel, Vashir)
1b. Shrubs or trees; flowers not as above 2
2a. Trees; flowers in panicles 3
2b. Shrubs; flowers in cymes, spikes or rarely panicles 5
3a. Corolla regular; stamens as many as corolla lobes 4
3b. Corolla 2 – lipped; stamens didynamous .... *Gmelina arborea* (Gambhari)
4a. Calyx not enlarged in fruit; corolla blue, purple or red .... *Callicarpa tomentosa* (Yesar)
4b. Calyx much enlarged and enclosing the fruit; corolla white .... *Tectona grandis* (Sak, Sagwan)
5a. Leaves simple 6
5b. Leaves compound, 3-5- foliolate 13
6a. Flowers often of two colours, in dense corymbose spikes .... *Lantana camara* (Ghaneri)
6b. Flowers of one colour, in cymose inflorescence 7
7a. Cymes lax, often few- flowered 8
7b. Cymes dense, many flowered .... *Avicennia officinalis* (Tuvarak)
8a. Flowers 5-8 mm long, greenish yellow, white or purplish 9
8b. Flowers more than 1 cm. long variously coloured 10
9a. Scandent shrubs; flowers purplish; calyx truncate .... *Premna herbacea* (Bharang)
9b. Erect shrubs or a small tree; flowers greenish – yellow; calyx 2 lipped .... *Premna serratifolia* (Agnimanth)
10a. Flowers white or with pinkish tinge 11
10b. Flowers blue .... *Clerodendrum serratum* (Bhoomijambu)
11a. Cymes lax, few flowered 12
11b. Cymes dense, many flowered .... *Clerodendrum multiflorum*
12a. Plants glabrous .... *Clerodendrum inerme* (Vamayuthika)
12b. Plants silky pubescent .... *Clerodendrum viscosum* (Kari, Bhandir)

13a. Leaflets 5-7, lanceolate, 0.5- 2 cm wide, greyish hairy beneath .... *Vitex agnus –castus* (Renuka)

13b. Leaflets 3-5, elliptic – ovate to lanceolate, 1.5 – 3.5 cm wide, thinly hairy beneath .... *Vitex negundo* (Nirgundi)

131. FAMILY – Violaceae (KAMAPUSHPA Kul.)

1a. Plants wild; stem branched; leaves elliptic – lanceolate; flowers deep rosy-red .... *Hybanthus enneaspermus* (Purushratna)

1b. Plants cultivated; stem very short or absent; leaves ovate – cordate; flowers deep violet, rose or white .... *Viola odorata* (Kamapushpa)

132. FAMILY – Vitaceae (DRAKSHA Kul.)

1a. Tendril climbers or bushy herbs 2

1b. Erect shrubs or undershrubs 6

2a. Plants cultivated for fruits produced in large hanging bunches .... *Vitis vinifera* (Draksha)

2b. Plants not as above 3

3a. Stem and branches 4- angled .... *Cissus quadrangula* (Asthisandhan)

3b. Stem and branches not angled 4

4a. Leaves 3- foliolate 5

4b. Leaves simple; tendrils often absent .... *Cissus adnata* (Naden)

5a. Leaves covered with hard bristles, thick; berries red .... *Cissus setosa* (Khajgoli)

5b. Leaves hairy; berries brown .... *Cayratia trifolia* (Amlaparni)

6a. Leaves simple, large, ovate- cordate .... *Leea macrophylla* (Dholsamudrika)

6b. Leaves pinnate 7

7a. Leaves simply pinnate .... *Leea crispa* (Bunchel)

7b. Leaves 2-3 pinnate .... *Leea sambucina* (Karkani)

133. FAMILY – Zingiberaceae (HARIDRA Kul.)

1a. Leaves spirally arranged; calyx red, stiff; flowers white .... *Costus speciosus* (Kemuk)

1b. Leaves distichous; calyx not as above; flowers variously coloured 2

2a. Lateral staminodes broad 3

2b. Lateral staminodes narrow or absent 9

3a. Flowers in dense spikes with green flower bearing bracts below and coloured empty bracts at the top 4

3b. Spikes not as above 8

4a. Spikes arising from centre of the leaves 5

4b. Spikes arising laterally, away from the leaves 6

5a. Plants cultivated; root – tubers deep yellow inside .... *Curcuma longa* (Haridra)

5b. Plants wild; root- tubers pale, yellow inside . ... *Curcuma amada* (Aamraharidra)

6a. Root tubers sessile, yellow or grey inside 7

6b. Root tubers at the end of long fibres, white inside, not aromatic . ... *Curcuma angustifolia* (Tavakshir)

7a. Root tubers yellow inside, aromatic .... *Curcuma aromatica* (Vanaharidra)

7b. Root tubers grey inside, not aromatic .... *Curcuma caesia* (Krishna Haridra)

8a. Root stock with aromatic tubers; leaves two, 6-12 cm long broadly ovate, spike 6-12 flowered; flowers white .... *Kaempferia galanga* (Chandramulika)

8b. Root stock with non- aromatic tubers; leaves three, upto 45 cm long, oblong; spikes 4-6 flowered; flowers pink or lilac .... *Kaempferia rotunda* (Bhoomichampak)

9a. Flowers in lax – flowered spikes or panicles 10

9b. Flowers in dense – flowered, cone- like spikes 12

10a. Inflorescence leafless; capsules white marked with vertical fine ribs .... *Elettaria cardamomum* (Ela)

10b. Inflorescence terminating a leafly stem; capsules red 11

11a. Capsules about 1 cm in diameter .... *Alpinia galanga* (Koshta Kulanjan)

11b. Capsules smaller, less than 1 cm in diameter .... *Alpinia officinarum* (Kulanjan)

12a. Anther cells spreading on a short filament .... *Amomum aromaticum*

12b. Anther cells parallel 13

13a. Leaves glabrous, beneath; bracts green 14

13b. Leaves pubescent beneath; bracts reddish .... *Zingiber purpureum* (Vanardrak)

14a. Plants cultivated; rhizomes dull white inside, strongly aromatic .... *Zingiber officinale* (Ardrak)

14b. Plants wild; rhizomes yellow inside, faintly aromatic or not .... *Zingiber zerumbet* (Kulanj)

134. FAMILY – Zygophyllaceae (KSHUDRAGOKSHUR Kul.)

1a. Trees; flowers blue .... *Guaiacum officinale* (Lohakashtha, Chobehayat)

1b. Herbs or undershrubs; flowers yellow or pink 2

2a. Leaves pinnate; stipules not spinous;
flowers yellow; fruits spinous or winged 3

2b. Leaves 1-3- foliolate, stipules spinous;
flowers pink; fruits not spinous ....*Fagonia cretica* (Dhanwayas)

3a. Plants silky- villous; fruits winged .... *Tribulus alatus* (Hasak)

3b. Plants pubescent; fruits spinous .... *Tribulus terrestris* (Kshudragokshur)

5

FAMILY - WISE LIST OF COMMON INDIAN MEDICINAL PLANTS

An exhaustive list of plant species provides correct and up-to date scientific name together with names in English, Sanskrit, Marathi and Hindi languages and plant parts used in medicine.

No.	Family		Latin Name	English Name	Sanskrit Name	Marathi Name	Hindi Name	Parts Used
1.	Acanthaceae	1	*Acanthus ilicifolius* L.	Holy leaved acanthus	—	Marandi	Harkukant	Roots, leaves
		2	*Adhatoda zeylanica* Medic [*A. vasica* Nees]	Malabar nut	Vasa	Adulsa	Adusa	Leaves, flowers, roots
		3	*Andrographis paniculata* (Burm. F.) Wall ex Nees	Creat	Kalmegh	Kalmegh, Pale kirayat	Kalpanath	Entire plant
		4	*Barleria cristata* L.	—	Sahachar	Nili koranti	Piyavasa	Entire plant
		5	*Barleria prionitis* L.	—	Saireyak	Kate koranti	Katsaraiya	Entire plant
		6	*Blepharis edulis* Pers.	—	Uchatta	Utangan	Utangan	Seeds
		7	*Blepharis repens* (Vahl) Roth	—	—	Hadsan	Hadsan	Entire plant
		8	*Crossandra infundibuliformis* (L.) Nees	—	Saireyak	Aboli	Salkat Sarai	Roots and leaves
		9	*Eranthemum roseum* (Vahl) R.Br.	—	—	Dasamuli	Gulsham	Roots
		10	*Hygrophila schulli* (Buch. Ham.) M.R. and S.M. Almeida	—	Kokilaksha	Kolsunda	Talimkhana	Seeds
		11	*Justicia procumbens* L.	—	—	Ghati pitapapada	—	Entire plant
		12	*Rhinacanthus nasutus* (L.) Kurz	—	Yuthiparni or Yuthikparni	Gajkarn	Palakjuhi	Roots, seeds, leaves

No.	Family		Latin Name	English Name	Sanskrit Name	Marathi Name	Hindi Name	Parts Used
		13	*Rungia parviflora* Nees	Fumitory, Trailing Rungia	Paryatak	Pittapapada	Khadsalia, Pittapapda	Entire plant
2	Adiantaceae	1	*Adiantum capillus-veneris* L.	Maiden hair fern	Hansraj	Hanspadi	Samalpatti	Entire plant
		2	*Adiantum pedatum* L.	Maiden hair fern	Hansraj	Hanspadi	Samalpatti	Entire plant
		3	*Adiantum philippense* L.	Maiden fern	Hanspadi	Hanspadi, Hansraj	Hanspadi	Entire plant
3	Agaricaceae	1	*Psalliota campestris* (L.) Fries	Mushroom	Chhatrak	Alambi	Bhui chhatta	Entire plant
		2	*Psalliota ostreata* Jacq.	Touchwood Button & Oyster	—	Phanas alambi	—	Entire plant
4	Aizoaceae	1	*Trianthema portulacastrum* L.	—	Vasuk, Raktapushpa Shilitika	Vasu	Safed punarnava	Roots
5	Alangiaceae	1	*Alangium salviifolium* (L.f.) Wang [*A. lamarckii* Thw.]	Sage leaved alangium	Aadot	Ankol	Dhera	Roots, bark
6	Alliaceae	1	*Allium cepa* L.	Onion	Palandu	Kanda	Piyaj	Bulb, seeds
		2	*Allium sativum* L.	Garlic	Rason	Lasun	Lahsun	Bulb
7	Altingiaceae	1	*Liquidambar orientalis* Mill.	Liquid storax	Silhak	Shilaras	Shilaras	Gum
8	Amaranthaceae	1	*Achyranthes aspera* L.	Prickly chaff flower	Apamarg	Aaghada	Latira	Entire plant seed-salt
		2	*Aerva lantana* Juss,	—	Shatakabhedi, Goraksh-agajja	Kapuri madhuri	Gorakhbuti	Roots, leaves, flowers
		3	*Amaranthus paniculatus* L.	Grain amranth	—	Rajgira	Bagi rajgiro	Flowers, seeds
		4	*Amaranthus roxburghianus* Nevsaki	—	Marsha, Bashpaka	Math, Tandulja	Marsa	Leaves
		5	*Amaranthus spinosus* L.	Prickly Amranth	Tandulia	Kante Math	Katailli cholai	Leaves
		6	*Amaranthus tricolor* L.	Amranth	Tandulia	Math	Lal Marsa	Leaves
		7	*Celosia argentea* L.	Cocks comb	Shitawar	Kardu	Sirimari, Safad murga	Roots, seeds
9	Amaryllidaceae	1	*Crinum asiaticum* L.	Poison bulb	Nagadmani	Gadanicha kanda, Nagadvana	Sukhdarshan	Bulb, leaves
		2	*Crinum latifolium* L.	—	—	Gadanicha kanda	Sukhdarshan	Bulb

No.	Family		Latin Name	English Name	Sanskrit Name	Marathi Name	Hindi Name	Parts Used
10	Anacardia-ceae	1	*Anacardium occidentale* L.	Cashew nut	—	Kaju	Kaju	Seeds
		2	*Buchanania lanzan* Spreng. [*B. latifolia* Roxb.]	Cuddapah almond	Priyal	Charoli	Chironji	Bark, seeds
		3	*Lannea coromandelica* (Houtt.) Merr. [*Odina woodier* Roxb.]	—	Gudmanjiri, Jingini	Shimpti, Moyi	Jingan	Bark, gum
		4	*Mangifera indica* L.	Mango	Aamra	Aamba	Aam	Fruits, seeds, leaves
		5	*Rhus parviflora* Roxb.	—	Tintidika	—	Samak-dana, Raitunga	Fruits
		6	*Semecarpus anacardium* L.f.	Marking nut Tree	Bhallatak	Bibba	Bhilawa	Fruits, seeds oil, bark
11	Annonaceae	1	*Annona reticulata* L.	Bullock's heart	Ramphal	Ramphal	Ramphal	Fruits
		2	*Annona squamosa* L.	Custard apple	Gandha-gatra	Sitaphal	Sitaphal, Sharifa	Fruits, leaves
		3	*Polyalthia longifolia* (Sonner.) Thw.	—	—	Ashok (Panancha)	Ashok	Bark
12	Apiaceae (Umbelliferae)	1	*Anethum graveolens* L.	Dill.	Shatapu-shpa, Shatava	Balanta-shepu	Sova, Bansoff	Fruits, leaves
		2	*Angelica glauca* Edgew.	—	Chorak	Chorova	—	Roots
		3	*Apium graveolens* L.	Celery	Aajmoda	Aajmoda, Koranja	Aajamoda	Leaves, fruits
		4	*Carum carvi* L. [*C. copticum* Hiern.]	Black caaraway	Krishnaj-irak	Shahjire, Kale jire	Kaljira	Fruits
		5	*Centella asiatica* (L.)Urb, [*Hydrocotyl asiatica* L.]	Indian pennywort	Manduk-parni	Karivana	Brahmi-bhed, Brahmi	Leaves
		6	*Coriandrum sativum* L.	Coriander	Dhanyak	Kothimbir, Dhane	Dhania	Leaves, Seeds
		7	*Cuminum cyminum* L.	Cumin	Jirak	Jire	Jira	Seeds
		8	*Daucus carota* L. var. *sativa* DC.	Carrot	Grunjnak	Gajar	Gajar	Roots
		9	*Ferula asafoetida* L.	Asafoetida	Hingu	Hing	Hing	Latex
		10	*Foeniculum vulgare* Mill.	Fennel	Mishreya	Badishep	Souff, Badisouff	Fruits
		11	*Trachyspermum ammi* (L.) Sprague	Bishop's weed	Yamanika, Yavani	Ova	Ajwain	Fruits
13	Apocynaceae	1	*Alstonia scholaris* (L.) R. Br.	Dita bark	Saptaparna	Satvin, Satvan	Chitvat	Bark

No.	Family		Latin Name	English Name	Sanskrit Name	Marathi Name	Hindi Name	Parts Used
		2	*Carissa congesta* Wight	Bengal currants	Karmardak•	Karvand	Karronda	Fruits
		3	*Carissa opaca* Stapf ex Haines	—	Karmardak	Karvand	Karronda	Fruits
		4	*Catharanthus roseus* (L.) G. Don	—	—	Sadaphuli	Sadabahar	Entire plant
		5	*Holarrhena pubescens* (Buch-Ham.) Wall. Ex G. Don [*H.antidysenterica* Wall.]	Telicherry bark	Kutaj	Kuda	Korya, Kurrya	Bark, Seeds
		6	*Inchnocarpus frutescens* (L.) Ait.	—	Ananta	Shamlata	Kali Anantmul	Roots
		7	*Nerium oleander* L. [*Nerium indicum* Mill; *N. odorum* Soland.]	Sweet scented oleander	Karvir	Kaner	Kaner, Kaniel	Roots, root-bark
		8	*Plumeria rubra* L.	Oleander	Kshircha-mpak	Khurchapha	Gulchit	Bark, leaves, flowers
		9	*Rauvolfia serpentina* Benth.	Serpina	Sarpaga-ndha	Hadki, Adkai	Dhaval-barua	Roots
		10	*Tabernaemontana divaricata* (L.) R. Br.	Wax flower	Nadyavarta, Tagar	Tagar	Tagar	Roots, latex
		11	*Thevetia peruviana* (Pers.) K. Schum. [*T. neriifolia* Juse ex Steud.]	Exile tree, Lucky nut	Pitkarvir	Pivali kaner	Pila kaner	Bark
14	Araceae	1	*Acorus calamus* L.	Sweet flag	Wacha, Ugragandha	Wekhanda	Bach, Ghorbach	Rhizome
		2	*Alocasia macrorhiza* (L.) G. Don	Giant taro	Manak, Mahapatra	Kansalu	Mankanda	Rhizome, Leaves
		3	*Amorphophallus bulbifer* Bl.	—	Suran	Suran	Surankand	Corm
		4	*Amorphophallus paeoniifolius* Bl.	Elephant's foot	Suran, Arshohan	Suran	Sukhkanda, Jimikand	Corm
		5	*Amorphophallus sylvaticus* (Roxb.) Kunth	—	Vajrakand, Aranyasuran	Jungli suran	—	Corm
		6	*Colocasia esculenta* (L.) Schott [*C. antiquorum* Schott]	Vegetable Arum	Kachvi	Alu	Arui, Kachu	Rhizome. Leaves
15	Arecaceae	1	*Areca cathecu* L.	Betle nut palm	Pugphal, Pugi	Supari, Pophal	Supari	Seed, fruit
		2	*Borassus flabellifer* L.	Palmyra palm, Brab tree	Tal	Tad	Tad. Tar	Stem-juice, Fruits

No.	Family		Latin Name	English Name	Sanskrit Name	Marathi Name	Hindi Name	Parts Used
		3	*Calamus rotung* L.	Cane palm	Wetra	Wet	—	Fruits
		4	*Caryota urens* L.	Fish tail, Palm	—	Surmad	—	Fruits
		5	*Cocous nucifera* L.	Coconut	Narikel	Naral, Mad	Narial	All parts
		6	*Daemonorops kurzianus* Becc. & Hook. f.	Dragon's blood	Raktaniryas	Hiradakhan	Khunkha-raba	Fruit-juice
		7	*Lodoicia maldivica* (Poir.) Pers.	Double cocount	Arkrag, agnijar, Abdhinarikel	Dariyacha naral	—	Endosperm
		8	*Phoenix dactylifera* L.	Date palm	Kharjuri, Pindkharjur	Khajur, Kharik	Khajur	Fruits
		9	*Phoenix sylvestris* (L.) Roxb.	Toddy palm of India	Vit Kharjur	Shindi	Sindi, Sendhi	Stem-juice, Fruits
16	Aristolochi-aceae	1	*Aristolochia bracteolata* Lamk. [*A. bracteata* Retz.]	Worm seed, Santonica, Bracteated birth wort	Dhoomra-patra	Gindhan	Kidamar, Gandhani	Seeds
		2	*Aristolochia indica* L.	Indian birth wort	Ishwari	Sapsand, Sapsan	Isarmool, Ishwarmool	Entire plant
17	Asclepiada-ceae	1	*Calotropis gigantea* (L.) R. Br.	Gigantic swallow wort, Madar	Shwetark	Mandar, Pandhari rui	Aak, Mandar	Roots, leaves
		2	*Calotropis procera* (Ait.) R. Br.	Mandar	Raktark	Rui	Aak	Roots, leaves
		3	*Ceropegia bulbosa* Roxb.	—	Patal tumbadi	Khaparkadu	—	Root-tuber
		4	*Gymnema sylvestre* (Retz.) R. Br. Ex. Schult.	—	Meshshr-ungi	Medshing, Bedki	Gudmar	Roots, leaves
		5	*Holostemma adakodien* Schult. [*H. rheedianum* auct. non Spreng.]	—	Arkpushpi	Khandodka. Shirdodi, Tultuli	Moranada, Chharivel	Roots, leaves
		6	*Leptadenia reticulata* (Retz.) Wt. & Arn.	—	Jivanti	Khandodki	Jivanti, Dodi	Roots, leaves
		7	*Marsdenia tenacissima* Wt. & Arn.	—	Moorwa, Madhurasa	Moorwa	—	Entire plant
		8	*Sarcostemma acidum* (Roxb.) Voigt. [*S. brevisiigma* Wt. & Arn.]	—	Somlata	Ransher	Somlata	Entire plant
		9	*Tylophora indica* (Burm. f.) Merr.	—	Pittavalli, Mulrasana	Pittamari, Khadakirasna	Antamul	Leaves, roots
18	Asteraceae (Compositae)	1	*Anacyclus pyrethrum* DC.	—	Aakarkarabl	Aakkalkara	Akarkara	Roots

No.	Family	Latin Name	English Name	Sanskrit Name	Marathi Name	Hindi Name	Parts Used
	2	*Artemisia maritima* L.	Worm seed	Kitmari, Yamani chohar	Kirmaniowa	Kirmani ajwan	Fruits
	3	*Artemisia nilagrica* (Cl.) Pomp. [*A. vulgaris* auct. Non L.]	Indian wormwood	Survarna	Surpin	Matjari	Entire plant
	4	*Blumea lacera* (L.) DC.	Dogbush	Kukundhar, Sukshm-apatra	Kukurbanda	Kukronda	Entire plant
	5	*Carthamus tinctorius* L.	Safflower	Kusumbha, Vastraranjak	Kardai	Kusumbha	Leaves, inflorescenc, fruits
	6	*Centipeda minima* (L.) A. Br. & Aschers	Sneeze wort	Chhikkika	Naakshin-kani	Chhikani	Fruits, leaves
	7	*Centrantherum anthelminticum* (L.) O. Ktze.	Purple fleabane	Aranyajirak	Kadujire	Kalajira, Somraj	Fruits
	8	*Cichorium intybus* L.	Endive, Wild chicory	Kasni	Kasni	Hinduba	Roots, leaves, fruits
	9	*Dendrathema indica* (L.) Des.	Chrysanthe mum	Shatpatri	Shevanti	Guldaudi	Roots, flowers
	10	*Echinops echinatus* L.	—	Utkantak	Utaktara, Kantechu-buk	Utkaatra	Inflorescence
	11	*Eclipta prostrata* (L.) L.	—	Bhrungraj	Maka	Bhangra	Entire plant
	12	*Elephantopus scaber* L.	—	Gojihva	Gojibh, Pathari	Gobhi	Entire plant
	13	*Emilia sonchifolia* (L.) DC.	—	Sadmandi	Sadmandi	Sdimodi	Entire plant
	14	*Eupatorium triplinerve* Vahl	Ayapana	—	Aypan	Ayapana	Leaves
	15	*Glassocardia basvallea* (L.f.) DC. [*G. linearifolia* Cass.]	—	—	Puneri pittapapada	Patha-rsuwa	Entire plant
	16	*Grangea maderaspatana* (L.) Poir.	—	Machipatri	Machipatra,	Massaru	Fruits
	17	*Guizotia abyssinica* L.	—	—	Korte, Karle	Ramtil	Fruits
	18	*Inula racemosa* Hook, f.	Orris root	Pushkarmul	Pushkarmul	Pohkarmul	Roots
	19	*Launaea procum bens* (Roxb.) Ramayya & Raj. [*I. nudicaulis* auct.]	—	—	Pathari	—	Leaves
	20	*Matricaria chamomilla* L.	Chamomil	—	Chamomil	Babuna	Oil
	21	*Phuchea lanceolata* (A.DC.) Oliver	—	Rasna, Yuktaaras	Rasna	Ramsum, Roshana	Leaves

No.	Family		Latin Name	English Name	Sanskrit Name	Marathi Name	Hindi Name	Parts Used
		22	*Saussurea lappa* Clarke	Costus, Kutroot	Kushtha	Uaplet, Koshtha	Kuth	Roots
		23	*Sphaeranthus indicus* L.	—	Mundi	Gorakh-mundi	Gorakh-mundi	Entire plant
		24	*Spilanthes oleracea* L.	—	Aakark-arabh	Aakkal-khada	Aakhal-khada	Inflorescence
		25	*Tagetes ereeta* L.	Big Marigold	Zendu	Zendu, Maskhmal	Genda	Leaves, flowers
		26	*Taraxacum officinale* Weber ex. Wigg.	Common dandelion	Dugdhpheni	Bathur	Dudhal, Dudhali	Roots
		27	*Tridax procumbens* L.	—	—	Dagdipala	—	Leaves
		28	*Vernonia cinerea* (L.) Less.	Purple fleabane	Sahadevi	Sahadevi, Sadodi	Sahadei	Entire plant
		29	*Wedelia calendulacea* Less.	—	Peet-bhrungaraj	Pivala maka	—	Entire plant
19	Averrhoaceae	1	*Averrhoa carambola* L.	Chinese gooseberry	Karmaranga	Karmal	Kamrak	Fruits
20	Balanitaceae	1	*Balanites aegyptia* (L.) Del. [*B. roxburghii* Planch.]	—	Endgudi	Higanbet	Hinguia, Hingot	Bark, fruits, seeds
21	Bambusaceae	1	*Bambusa arundinacea* (Retz.) Willd.	Thorny Bamboo	Wansh	Bambu	Bans	Gum, roots, leaves, seeds
22	Berberidaceae	1	*Berberis aristata* DC.	Berberry, Tree turmeric	Daruharidra	Daruhalad	Daruhaldi	Roots, wood, Gum
		2	*Podophyllum hexandrum* Royle	*Podophyllum*, May-apple	Giriparpat, Vanvruntak	Papara	Papara, Papri	Roots
23	Betulaceae (Corylaceae)	1	*Betula alnoides* Buch. Ham. [*Betula utilis* D. Don.]	Himalayan silver birch	Bhurjapatra, Bhurja	Bhurjapatra	Bhujpattra	Leaves
24	Bignoniaceae	1	*Dolichandrone falcata* (Wall. Ex. DC.) Seem var. *lawii* (Wall) Haines	—	meshshr-ungi	Medhshingi	Medhasingi	Leaves
		2	*Oroxylum indicum* (L.) Vent	—	Shyonak	Tentu	Shonak	Roots, bark
		3	*Stereospermum chelenoides* (L.f.) DC.	—	Patla, Tamrapushpi	Padal	Paral	Roots, bark, flowers
		4	*Stereospermum colais* (Buch Ham. Ex Dillw.) Mabb.	Trumpet flower	Shwet patla	Padal	Safed paral	Roots, flowers
		5	*Tecomella undulata* (Sm.) Seem	Wave leaved Bignonia	Rohitak	Rohida	Roheda	Bark

No.	Family		Latin Name	English Name	Sanskrit Name	Marathi Name	Hindi Name	Parts Used
25	Bixaceae	1	*Bixa orellana* L.	—	Sinduri	Shendri	Sandurema	Seeds
26	Bombacaceae	1	*Adansonia digitata* L.	Baobad, Monkey bread	Gorakshi, Ravanamlika	Gorakhch-inch	Gorakha-mali	Fruits
		2	*Bombax ceiba* L. [*B. malabaricum* DC.]	Silk cotton tree	Shalmali	Katesavar	Semal	Root, tubers, flowers, gum.
		3	*Ceiba pentandra* L.	White silk cotton tree	Kutshalmali	Pandheri savar	Safed semel	Root, leaves, gum
27	Boraginaceae	1	*Heliotropium indicum* L.	—	Hastis-hunda	Bhurundi	Hathishur	Entire plant
		2	*Onosma bracteatum* Wall.	—	Vrush jivha	Gavjaban	Gavjaban	Entire plant
		3	*Trichodesma indicum* (L.) R. Br.	—	Adhopushpi	Adhopushpi	Undahuli	Entire plant
28	Brassicaceae (Cruciferae)	1	*Brassica alba* Rabenh.	White mustard	Shweta rajika	Pandheri mohari	Safed rai	Seeds
		2	*Beassica juncea* (L.) Zern. et Coss.	Mustard	Rajika	Mohari	Rai, Lal rai	Seeds
		3	*Brassica napus* L.	Rape-seed	Shweta ragika	Pandhri mohari	Safed rai	Seeds
		4	*Brassica nigra* (L.) Koch	Black mustard	Rajika	Kali Mohari	Rai	Seeds
		5	*Lepidium sativum* L.	Garden cress	Chandr-ashur	Ahaliv, Aaliv	Halim, Chansur	Seeds
		6	*Raphanus sativus* L.	Radish	Mulak	Mula	Muli	Roots, fruits
		7	*Sisymbrium irio* L.	Hedge mustard	—	Rantikhi	Khaksi	Seeds
29	Bromeliaceae	1	*Ananas comosus* (L.) Merr.	Pineapple	—	Aananas	Ananas	Fruits
30	Burseraceae	1	*Boswellia serrata* Roxb.ex Coleb.	Indian polibanum	Kunduru	Salai	Salai	Gum
		2	*Commiphora myrrha* Engl.	Myrrh	Bol, Gandhras	Hirabol	Hirabol	Gum
		3	*Commiphora wightii* (Arn.) Bhandri [*C. mukul* (Hook ex Stocks) Engl.]	Bdelium	Guggul	Guggul	Guggul, Gugal	Gum
31	Cactaceae	1	*Opuntia elatior* Mill. [*O. dillenii* Grah.]	Prickly pear	Kanthari, Kanthasev	Phadya nivdung	Nagphani	All parts
		2	*Nopalea cochenillifera* (L.) Salm Dyck	Cochinel cactus	Kanthari	Phadya nivdung	Nagphani	All parts
32	Caesalpi-niaceae	1	*Bauhinia racemosa* Lamk.	Bidi leaf tree	Yamal-patrak, Ashmantak	Aapta	Sirhata	Bark, leaves
		2	*Bauhinia variegata* L.	Mountain ebony	Kanchnar, Kovidar	Kanchan	Kachnar	Bark

No.	Family		Latin Name	English Name	Sanskrit Name	Marathi Name	Hindi Name	Parts Used
		3	*Caesalpinia bonduc* (L.) Roxb. [*C. bonolucella* Flemm.]	Bonduc nut	Puti karanj, Kantaki karanj	Sagargota, Gajaga	Kanga, Kanta karanj	Roots, leaves, seeds
		4	*Caesalpinia decapetala* (Roth) Alston	—	—	Chillar	—	Seeds
		5	*Caesalpinia pulcherrima* (L.) Sw.	—	Krishan-achuda	Sankar-shawar	Gul-e-Tura	All parts
		6	*Caesalpinia sappan* L.	Sappan wood	Patadga	Patang	Patadga	Wood
		7	*Cassia absus* L.	—	Vannakulth	Chinnol Kankudi	Chakas	Seeds
		8	*Cassia auriculata* L.	—	Avartaki	Tarwat, Tarota	Tarwad	All parts
		9	*Cassia fistula* L.	Indian lanburnum	Aragwadh	Bahwa	Amaltash	Roots, leaves, pods
		10	*Cassia obtusifolia* L.	Foetid cassia	Chakramard	Takla	Chakwad	All parts
		11	*Cassia occidentalis* L.	Rubbish cassia	Kasmard	Kaswinda	Kasaundi	All parts
		12	*Cassia senna* L. [*Cassia angustifolia* Vahl.]	Indian senna	Markandika, Swarnpatri	Sonamukhi	Sonamaki	Leaves, pods
		13	*Cassia sophera* L.	Rubbish cassia	Kasmard	Kala Kaswinda	Kasaundi	All parts
		14	*Cassia tora* L.	Foetid cassia	Chakramard Prapunnat	Takla	Chakwad	All parts
		15	*Mouliava spicata* (Dalz.) Nicols. [*Wagatea spicata* Dalz]	—	Guccha karanj	Wakeri	—	Roots
		16	*Saraca asoca* de Willd.	Asoka	Ashoka, Hemapushpa	Ashok (Sitecha)	Ashok	Bark
		17	*Tamarindus indica* L.	Tamarind	Amlika	Chincha	Imli, Amli	All parts.
33	Cannabaceae	1	*Cannabis sativa* L.	Indian hemp	Bhanga	Bhang	Bhang	Leaves, inflorescence, seeds
34	Capparida-ceae	1	*Cadaba fruticosa* (L.) Druce [*C. indica* Lamk.]	—	—	—	Kodhan, Habab	Roots, leaves
		2	*Capparis decidua* (Forsk.) Edgew. [*C. aphylla* Roth.]	—	Karir, Granthil	Nepati, Sedad	Karir, Karil	Bark, buds, fruits
		3	*Capparis sepiaria* L.	Indian Caper	Kakadani, Hinsra	Grudhra-nakhi	Kanthari	Fruits.
		4	*Capparis zeylanica* L.	—	Vyaghra-nakhi	Govund-aphal,	Bagnaha, Giloran	Roots, fruits.

No.	Family		Latin Name	English Name	Sanskrit Name	Marathi Name	Hindi Name	Parts Used
		5	*Crateva adansonii* DC. subsp. *odora* (Buch. Ham.) Jacobs. [*C. murvala* Buch. Ham.] [*C. religiosa* auct. Non Forsk.]	Holy garter, Pear	Ashmaari-ghna Varun	Vayavarna, Varun	Barun	Bark, leaves
35	Caricaceae	1	*Carica papaya* L.	Papaw, Papaya	Erandkaar-kati Madhu-karkati	Popai	Papita	Leaves, fruits latex.
36	Cassythaceae	1	*Cassytha filiformis* L.	Doddar	Akashalli	Akashvel	Amarvel	Entire plant
37	Celastraceae	1	*Cassine albens* (Retz.) Kosterm [*C. glauca* (Rottb.) O. ktze.]		—	Tamraj, bhutakeshi	Chori	Roots, bark, leaves
		2	*Celastrus paniculata* Willd.	Staff tree	Jyotishmati	Malkangoni	Malkangni	Seeds, seed-oil
		3	*Maytenus emarginata* (Willd.) Ding Hou	—	—	Hekal	Hekal	Leaves, fruits
38	Ceratophy-llaceae	1	*Ceratophyllum demersum* L.	—	Shaival	—	—	Entire plant
39	Chenopod-iaceae	1	*Basella alba* L. var. *alba* var. *rubra* (L.) Stew.	Indian Spinach	Upodaki	Mayalu	Poy, Poie	Leaves
		2	*Beta vulgaris* L. var. *rapa*. Dum	Beet Root	Raktag-runjnak	Beet	Beet, Chukandar	Roots
		3	*Chenopodium album* L.	Lamb's quarters	Vastuk, Yavshak	Chakvat	Bathua	Stem, leaves
		4	*Chenopodium ambrossioides* L.	—	Sugandha, Vastuk	Chakvat	—	Stem, leaves
		5	*Chenopodium murale* L.	—	—	Chakvat	—	Stem, leaves
		6	*Spinacia oleracea* L.	Spinach	Palakya	Palak	Palak	Stem, leaves
40	Cleomaceae	1	*Cleome chelidonii* L. f.	—	—	—	—	Leaves
		2	*Cleome gynandra* L.	—	Bastagan-dha.	Tilvan	Hurhur	Roots, seeds
		3	*Cleome viscosa* L.	—	Adityabhakta	Pevali Tilvan	Hurhur	Leaves, seeds
41	Clusiaceae (Guttiferae)	1	*Calophyllum inophyllum* L.	Pannay tree	Punnag	Undi	Kala nagkshar	Flowers, buds
		2	*Garcinia cambogia* (Gaertn.) Desr.	—	—	—	—	Fruits
		3	*Garcinia indica* Choisy	Kokam tree	Vruk shamla, Saramla	Kokam, Ratamba	Kokam	Bark, fruits, seed-oil
		4	*Garcinia morella* (Gaertn.) Desr.	Ceylon gamboge	Tamal	Tamal	Tamal	Gum

No.	Family		Latin Name	English Name	Sanskrit Name	Marathi Name	Hindi Name	Parts Used
		5	*Garcinia pedunculata* Roxb.	—	Aamlawatas	Amlawatas	Amarvel, Thikla	Fruits
		6	*Mammea longifolia* (Wt.) Planch.	Alexandrian laurel	Surpunnang Nameru	Surangi	Lal nagkesar	Flowers, buds
		7	*Mesua nagassaricum* (Burm. f.) Koster [*Mesua ferrea* auct non L.]	Cobra's saffron	Nagkeshar	Nagkeshar	Asali Nagkesar	Stamens
42	Cochlospermaceae	1	*Cochlospermum religiosum* (L.) Alston	Yellow cotton tree	Peetkarpas	Galgal	Pili kapas	Leaves, seeds, gum
43	Combretaceae	1	*Anogeissus latifolia* (Roxb. Ex. DC.) Wall. Ex. Guill. & Perr.	Axel wood, Button tree	Dhav, Dhurandhar	Dhavada, Dhamoda	Dhaora, Dhav	Gum
		2	*Combretum albidum* G. Don.	—	—	Haladvel	—	—
		3	*Quisqualis indica* L.	Rangoon Creeper	—	Rangunvel	—	Seeds
		4	*Terminalia alata* Heyne ex Roth. [*T. tomentosa* (DC.)Wt. & Arn]	—	Aasan	Aain	Asan	Bark
		5	*Terminalia arjuna* (Roxb.) Wt & Arn.	Arjuna, Myrobalans	Arjun	Arjun Sadada, Arjun	Arjun, Kahua	Bark
		6	*Terminalia bellirica* (Gaertn.) Roxb	Myrobalans, Bedda nut, Chebulic	Bibhitak	Behada	Baheda	Fruits
		7	*Terminalia chebula* Retz.	Myrobalans	Haritaki, Abhaya	Hirda, Harda, Balhirda	Harad, Harar	Fruits, infected leaves
44	Convolvulaceae	1	*Argyreia nervosa* (Burm. f.) Boj. [*A.speciosa* Sm.]	Elephant creeper	Vruddhadarak	Samudrashok	Samandarsokh	Roots, leaves
		2	*Convolvulus arvensis* L.	—	—	—	Hiranpad	Roots
		3	*Convolvulus microphyllus* Sieb. Ex. Spreng	—	Shankhapushpi	Sankhwel, Shankhahuli	Shankhpushpi	All parts
		4	*Convolvulus pluricaulis* Choisy	—	Shankhpushpi	Shankhawali	Shankhpushpi	All parts
		5	*Cressa cretica* L.	—	Rudanti, Chanakpatri	Ranhaarbhara, Khardi	Rudravanti	All parts
		6	*Evolvulus alsinoides* (L.) L.	—	Vishnukranta	Nili shankhawali	Bhurishankhahuli	All parts
		7	*Ipomoea alba* L.	Moon flower	—	Chandrakant, Chandani	Gulchandani	All parts, seeds

No.	Family		Latin Name	English Name	Sanskrit Name	Marathi Name	Hindi Name	Parts Used
		8	*Ipomoea aquatica* Forsk.	Swamp cabbage	Kalambi, Nadi	Nalachi bhaji	Kalmisag	Stem, leaves
		9	*Ipomoea batatas* (L.) Lamk.	Sweet potato	—	Ratale	Sakarnand	Root-tuber
		10	*Ipomoea mauritiana* Jacq. [*I. digitata* auct. non L.]	—	Vadaarikand, kahirvidari	Vidarikand	Balaikand	Tubers
		11	*Ipomoea nil* (L.) Roth [*I. hederacea* auct. non Jacq.]	—	Shankhini	Kaladana	Mirche	Seeds
		12	*Ipomoea pes-caprae* (L.) Sweet [*I. biloba* Forsk.]	Sand binder	Maryadvalli, Vruddha-darak	Maryadvel	Dopatilata	Roots, leaves
		13	*Ipomoea sepiaria* Koen. ex. Roxb.	—	Lakshmana	Laxmana	Laxmana	Roots
		14	*Merremia emarginata* (Burm.f.) Hall. f. [*Ipomoea reniformis* Choisy]	—	Musha-kkarni	Undirkani	Musakani, Chuhakani	All parts
		15	*Operculina turpethum* (L.) Silva Mansoa [*Ipomoea turpethum* (L.) R.Br.]	Indian jalup	Trivrut, Nishottar	Nishottar, Shetvad	Nishot	Roots
		16	*Rivea laotica* Ooststr. [*R. ornata* (L.) Peers.]	—	Funjika	Fand	Fand	Leaves, plant-juice
45	Crassulaceae	1	*Kalanchoe pinnata* (Lamk.) Pers. [*Bryophyllum calycinum* Sslisb.]	—	Parnabij	Ghaimari, Panphuti	Dardmar, Jakhm-ehayat	Leaves
		2	*Kalanchoe laciniata* (L.) Pers.	—	Parnabij	Panphuti	Dardmar	Leaves
46	Cucurbit-aceae	1	*Benincasa hispida* Thunb. [*B. cerifera* Savil].	Ash pumpkin, Ashguard	Kushmandh	Kohala	Petha, Ehura, Khumhada	Fruits
		2	*Citrullus colocynthis* (L.) Schrad.	Colocynth, Bitter gourd	Indravaruni, Vishala	Indrayan	Indrayan	Roots, fruits
		3	*Citrullus lanatus* (Thunb) Mustet. Nakai	Water-melon	Kalingad	Kalingad, Tarbuj	Tarbuz	Seeds
		4	*Coccinia grandis* (L.) Voigt. [*C. cordifolia* Cogn.]	Ivyguard	Raktaphala, Tundikeri	Tondali, Tondale	Kanduri, Kunali	All parts
		5	*Corallocarpus epigaeus* (Rottl. & Willd.) Hook. f.	—	Katunahi	Garajphal	Rakasgatta	Root-tuber
		6	*Cucumis callosus* (Rottl.) Cogn.	—	—	Shendani	Karit	Fruits

No.	Family	Latin Name	English Name	Sanskrit Name	Marathi Name	Hindi Name	Parts Used
7		*Cucumis melo* L. var. *melo*	Melon	Kharbuja, Dashangul	Kharbuj	Kharbuja	Seeds
8		*Cucumis melo* var. *utilissimus* (Roxb.) Duthie & Fuller	Snake cucumber	Karkati, Erwaru	Kakdi	Kakri	Fruits, seeds
9		*Cucumis sativus* L.	Cucumber	Trapus	Tause	Khira	Fruits, seeds
10		*Cucurbita maxima* Duch. Ex Lamk.	Red pumpkin	Rakta kushmand	Tambada bhopala	Kaddu, Lal kumhada	Fruits, seeds
11		*Cucurbita moschata* (Duch. Ex Lamk.) Duch. Ex. Poir	Pumpkin	—	Bhopala, Kashiphal	Kumhada	Fruits, seeds
12		*Cucurbita pepo* L.	Field pumpkin	Kushmandi, Karkaru	Dangar	Safed Kaddu	Seeds
13		*Diplocyclos palmatus* (L.) Jeffery [*Bryonopsis laciniosa* Naud.]	—	Buj	Shivlingi	Bajgureia	Entire plant
14		*Lagenaria siceraria* (Molina) Standley [*L. vulgaris* Seringe]	Bottle gourd	Kututumbi	Kadu bhopala, Dudhya bhopala	Kadavit-umbi	Fruits, seeds
15		*Luffa acutangula* (L.) Roxb	Ridged gourd	Koshataki	Dodaka, Shirale	Torai	Roots, leaves
16		*Luffa amara* Roxb	—	Koshataki	Ran dodaka, Kaadu dodaka	Kadavi torai	Fruits
17		*Luffa cylindrica* (L.) Roem. [*I. aegyptiaca* Mill.]	Sponge gourd	Mahako-shataki	Ghosale	Baditorai	Leaves
18		*Luffa echinata* Roxb.	Bristly luffa	Devdali, Dhamargav	Devdangri, Kadughosale	Devdali, Nenua	All parts
19		*Melothria heterophylla* (Lour.) Cogn.	—	—	Gometi	Gometi, Tarali	All parts
20		*Melothria maderaspatana* (L.) Cogn.	—	Aahilekhan	Chirati	Aagumki, Nilari	Entire plant
21		*Momordica charantia* L.	Bitter gourd	Karvellak	Karale	Karela	All parts
22		*Momordica dioica* Roxb. Ex Willd.	—	Karkotaki	Kartoli	Kakola	All parts
23		*Trichosanthes anguina* L.	Snake-gourd	Chichinda	Padval	Chachinda	Fruits, seeds
24		*Trichosanthes cucumerina* L.	—	Patol	Ran padval	—	Roots, leaves, fruits
25		*Trichosansthes dioica* Roxb.	Wild snake gourd	Patol	Parvar	Parvar	Roots, leaves, fruits
26		*Trichosanthes tricuspidata* Lour.	—	Mahakal	Koundal	Lal Indrayan, Mahurar	Roots, fruits

No.	Family		Latin Name	English Name	Sanskrit Name	Marathi Name	Hindi Name	Parts Used
47	Cupressaceae	1	*Juniperus communis* L.	Juniper Berry	Hapusha, Ashwatth-aphal	Hosh	Hauber, Huber	Wood, fruits
48	Cyperaceae	1	*Cyperus rotundus* L.	Nut grass	Musta, Mustak	Motha	Motha, Musta	Root-tubers
		2	*Cyperus scariosus* R. Br.	Nut grass	Nagar mustak	Nagarmotha	Nagarmotha	Root-tubers
		3	*Kyllinga bulbosa* Beauv. [*K. truceps* Rottb.]	—	Nirvish	Nirvishi	Nirvis	Roots
		4	*Scirpus kysoor* Roxb.	—	Kesaru	Kachara, Kaisargadda	Keseru	Rhizome
49	Dilleniaceae	1	*Dillenia indica* L.	Indian Dillenia	Bhavyam	karambal	Chalata	Bark, fruits
		2	*Dillenia pentagyna* L.	—	—	Lahan karambal	Chota chalta	Fruits
50	Dioscoreaceae	1	*Dioscorea alata* L.	Yam	Dandalu	Kachil	Ratalu	Bulbis, root-tuber
		2	*Dioscorea deltoidea* Wall.	—	—	—	—	Entire plant
		3	*Dioscorea esculenta* L.	—	—	Kanage	—	Root-tuber
		4	*Dioscorea bulbifera* L.	Wild yam	Warahi kand	Dukkar-kand, Kadu-karanda	—	Root-tuber
51	Dipteroca-rpaceae	1	*Dipterocarpus alatus* Roxb.	—	Ashwakarna Kashtha tail	Garjan	Garjan, Jayan	Wood-oil
		2	*Shorea robusta* Gaertn.	Sal tree	Sarj, Shal	Ralecha vruksha	Sal, Sakhua	Latex
		3	*Vateria indica* L.	—	Ajakarna, Marichp-atrak	Chandrus	Safed damar	Latex, seed-oil
52	Ebenaceae	1	*Diospyros exculpta* Buch. Ham.	Bidi leaf	Kak tinduk	Temburni	Tendu	Bark, fruits
		2	*Diospyros melanoxylon* Roxb.	Bidi leaf	Kak tindum	Temburni	Tendu	Bark, fruits
		3	*Diospyros malabarica* (Desr.) Kostel. [*Diospyros peregrina* auct] [*D. embryopteris* Pers.]	Indian persimon	Tinduk	Temburni	Tendu	Bark, fruits
53	Ehretiaceae	1	*Cordia dichotoma* Forst. F. [*C. myxa* auct, non L.]	Sebestan	Shleshm-antak, Bahuvar	Bhokar	Lisoda	Roots, bark
		2	*Cordia macleodii* (Griff.) Hook. f. & Thoms	—	—	Dahivan	—	Bark
		3	*Cordia monoica* Roxb.	—	—	Lahan bhokar	—	Fruits

No.	Family		Latin Name	English Name	Sanskrit Name	Marathi Name	Hindi Name	Parts Used
		4	*Cordia sinensis* Lamk [*C. rothii* Roem. & Schult]	—	—	Gondani	—	Fruits
		5	*Ehretia aspera* Willd	—	Ajanwru-ksha	Aajanvru-ksha	—	—
		6	*Ehretia laevis* Roxb	—	Ajanwru-ksha	Aajanvru-ksha	—	—
54	Ephedraceae	1	*Ephedra gerardiana* Wall.	—	Som	Somvalli	Assmanibuti	All parts
55	Euphorbi-aceae	1	*Acalypha indica* L.	Indian Acalypha	Haritm-anjiri	Khokli, Kupi	Kuppi	All parts
		2	*Aleuriles moluccana* Willd.	—	—	Jungali Aakrod	Jungali aakharod	Fruits, seed oil
		3	*Baliospermum solanii-folium* (Burm.f.) Suresh [*B. montanum* (Willd.) Muell. Arg.]	—	Danti	Danti	Danti	Roots
		4	*Bridelia crenulata* Roxb	—	—	—	—	Bark
		5	*Bridelia montana* (Roxb.) Willd.	—	Ekvir	Hasan, Aasana	Khaj, Kharak	Bark
		6	*Chrozophora prostrata* Dalz.	—	—	—	Pattahar phod	Roots
		7	*Chrozophora rottleri* (Geis) Juss. Ex Spreng.	—	—	—	Pattahar phod	Roots
		8	*Croton oblongifolius* Roxb.	—	Nagdanti	Ghansar	Chak, Hukum puter	Roots, bark, leaves
		9	*Croton tiglium* L.	Purging croton	Jaipal	Jaipal	Jamalgota	Seeds, seed-oil
		10	*Euphorbia antiquorum* L.	Tringular sperge	Tridhari, Sehund, Vajri	Tindhari nivdung	Tridhari thoher	Roots, stem
		11	*Euphorbia hirta* L.	—	Nagarjuni, dugdhika	Mothi Naiti	Dudhia	Entire plant
		12	*Euphorbia indica* Lamk. [*E. hypericifolia* L.]	—	—	Naiti	Hajardana	Entire plant
		13	*Euphorbia neriifolia* L.	Common milk hedge	Snuhi	Nivadung	Sehund, Thohar	All parts
		14	*Euphorbia nivulia* Buch. Ham.	—	Snuhi	Vainivdung	Sehund, Thoher	All parts
		15	*Euphorbia thymifolia* L.	—	Laghudu-gdhika	Lahan Naiti	Choti dudhi	All parts
		16	*Euphorbia tirucalli* L.	—	—	Sher	Thohar	Latex
		17	*Excoecaria agallocha* L.	—	—	Geva, Fungli	Hura	Bark, latex
		18	*Jatropha curcas* L.	—	—	Mogali erand	Moglai Erand	Roots, latex

No.	Family		Latin Name	English Name	Sanskrit Name	Marathi Name	Hindi Name	Parts Used
		19	*Jatropha gossipifolia* L. var. *elegans* DC.	—	—	Vilayati erand	—	Leaves
		20	*Mallotus philippensis* (Lamk.) Muell. Arg.	Monkey face tree	Kampillak	Kapila	Kapila, Kamila	Fruit-powder
		21	*Phyllanthus acidus* (L.) Skeels [*Cicca acida* L.]	Star goose-berry	Lavli, Sugandh-mula	Rai awla	Harfa rewdi	Fruits
		22	*Phyllanthus amarus* Schum. [*P. niruri* auct. Non L.]	Feather foil plant	Bhumya-malaki	Bhui awli	—	All parts
		23	*Phyllanthus emblica* L. [*Emblica officinalis* Gaertn.]	Emblic myrobalans	Amalaki	Awla	Amla	Fruits
		24	*Phyllanthus reticulatus* Poir.	—	Krishna kamboji	Puwan, Kamoni	Panjoli	Leaf-juice
		25	*Drypetes roxburghii* (Wall.) Hurus [*Putranjiva roxburghii* Wall.]	Indian amulet plant	Putranjeev	Putrajeev, Putjan	Jiyapota	Leaves, seeds
		26	*Ricinus communis* L.	Castor oil plant	Erand	Erandi	Arand, Rendi	Roots, leaves, seeds, seed-oil
		27	*Sebestiania chamaelea* (L.) Muell Arg.	—	—	Bhui erand	—	Seeds
		28	*Securinega leucopyrus* (Willd.) Muell. Arg.	—	—	Kante puwan	—	Leaf-juice
		29	*Tragia plukenetii* A. R. Sm.	—	Vruschikali	Kanchkuiri	Bavhanta	Roots
		30	*Trewia nudiflora* L.	—	Pindar	Siwani	Pindar, Bhilaur	Roots
56	Euryalaceae	1	*Euryale ferox* Salisb.	Gorgon fruit	Makha-nnam	Makhana	Makhana	Seeds
57	Fabaceae [Pappilion-aceae]	1	*Abrus precatorius* L.	Wild liquorice	Gunja, Raktika	Gunj	Gunch, Ratti	Roots, leaves
		2	*Alhagi pseudalhagi* (M.Bied) Desv.	Arabian manna plant	Yawas	Jawasi	Jawas, Hingua	All parts
		3	*Aarachis hypogaea* L.	Ground-nut	Bhushimbi, Mandapi	Bhuimug	Mungphalli	Seed, seed-oil
		4	*Butea monosperma* (Lamk.) Taub. [*B. frondosa* Koen. ex Roxb.]	Flame of the forest	Palash, Brahma vruksha	Palas	Palash, Dhak	Flowers, seeds, gum
		5	*Butea superba* Roxb.	—	Lata palash	Palas vel	Palash lata	Gum

No.	Family	Latin Name	English Name	Sanskrit Name	Marathi Name	Hindi Name	Parts Used
	6	*Cajanus cajan* (L.) Millsp. [*C. indicus* Spreng.]	Pigeon pea	Adhaki, Shanpus-hpika	Tur	Arhar	Leaves, seeds
	7	*Canavalia gladiata* (Jacq.) DC.	Sword bean	Kolshimbi	Abai, Fatadi	Bada sem	Seeds
	8	*Cicer arientinum* L.	Gram, Chick pea	Chanak	Chana, Harbhara	Bat, Chana	Leaves, seeds
	9	*Clitoria ternatea* L.	Winged leaved	Girikarnika, Aparajita	Gokarna, Kajali	Aprajit	Roots, flowers, seeds
	10	*Crotalaria juncea* L.	Sann hemp	Shan	Tag	San	Leaves, seeds
	11	*Crotalaria verrucosa* L.	—	Shanapu-shpi	Khulkhula	Shata-pushpi, Patsan	Leaves
	12	*Cyamopsis tetragonoloba* (L.) Taub. [*C. psoralioides* DC.]	Cluster bean	—	Gowar, Gawar	Gowar	Fruits
	13	*Dalbergia latifolia* Roxb.	Rose wood	Shinshapa	Shisaw	Shisam	Bark, leaves
	14	*Dalbergia sissoo* Roxb. Ex DC.	Black wood	Shinshapa	Shisaw	Shisam	Bark, leaves
	15	*Desmodium gangeticum* (L.) DC.	—	Shalparni, Deerghamula	Salvan, Ranmal	Salvan, Gauri	Roots
	16	*Desmodium oojeinensis* (Roxb.) Ohasi	—	Tinish	Tiwas	Sandan, Chanan	Bark, gum
	17	*Erythrina suberosa* Roxb.	Coral tree	Paribhadra	Pangara	Farhad	Bark, leaves
	18	*Erythrina variegata* L.	Coral tree	Paribhadra	Pangara	Farhed	Bark, leaves
	19	*Glycirrhiza glabra* L.	Liqourice	Yashtimadhu	Jyeshthi-madh	Mulhathi, Mulethi	Roots.
	20	*Indigofera tinctoria* L.	Indigo	Neelini	Neel	Neel	Roots, leaves
	21	*Lablab purpureus* (L.) Sweet var. *lignosus* (L.) King	Flat bean, Hyacinth bean	Nishpav, Rajashimbi	Vaal	Nishpav, Sem	Fruits, seeds
	22	*Lathyrus sativus* L.	Chickling Vetch	Kalay, Triput	Laang, Laakh	Khensari	Seeds
	23	*Macrotyloma uniflorum* (Lamk.) Verdc. [*Dolichos uniflorus* Lamk.]	Horse gram	Kulattha	Kulith, Hulga	Kalthi	Seeds
	24	*Melilotus indica* (L.) DC.	—	—	Ranmethi	—	Leaves
	25	*Mucuna pruriens* (L.) DC.	Cowitch plant	Kapikachhu Kawach beej	Khajkuhili	Kewanch, Kaunch	Bristles of fruit, roots, seeds
	26	*Paracalyx scariosa* (Roxb.) Ali [*Cylista scariosa* Roxb.]	—	—	Ranghewda	—	Roots

No.	Family	Latin Name	English Name	Sanskrit Name	Marathi Name	Hindi Name	Parts Used
	27	*Phaseolus vulgaris* L.	French bean	—	Shrawan ghewda	Sem	Fruits, seeds
	28	*Pisum sativum* L. subsp. *arvense* (L.) A. & G.	Pea	Harenu	Watana	Mattar	Seeds
	29	*Pisum sativum* L. subsp. *sativum*	Field pea	Sateen	Hirva vatana	Mutter	Seeds
	30	*Pongamia pinnata* (L.) Pierre [*P. glabra* Vent.]	Indian beech	Kaaranj, Naktamal	Karanj	Karanj, Kinnal	All parts, seed-oil
	31	*Psoralea corylifolia* L.	Psoralea seed	Bakuchi, Kushthaghni	Bavachi	Bakuchi	Seed-oil
	32	*Pterocarpus marsupium* L.	Malabar kino tree	Beejak, Asan	Bivala	Vijaysar, Bijesar	Wood, gum
	33	*Pterocarpus santalinus* L.	Red sandal-wood	Raktach-andan, Raktang	Raktach-andan	Raktch-andan	Wood
	34	*Pueraria tuberosa* (Roxb.ex Willd.) DC.	—	Widari kand	Bendar, Ghodvel	Siali	Root-tuber
	35	*Sesbania grandiflora* (L.) Poir.	—	Agastya, Agasti	Hadaga, Agasta	Agast, Hathiya	Roots, bark, leaves
	36	*Sesbania sesban* (L.) Merr. [*S. aegyptiaca* Poir.]	—	Jayanti, Jaya	Shewri	Jait	Bark, leaves, seeds
	37	*Tephrosia purpurea* (L.) Pers.	—	Shara-punkha	Unhali	Sarakonka	All parts
	38	*Teramnus labialis* (L.) Spr.	—	Mashparni	Ran udid	Mashaparni	Seeds
	39	*Uraria lagopoides* (Burm. f.) DC.	—	Prushni-parni	Pithwan	Pitwan	Roots
	40	*Uraria picta* (Jacq.) Desv. Ex. DC.	—	Prushni-parni	Pithwan	Dabra	All parts
	41	*Vigna aconitifolia* (Jacq.) Morechal	Moth bean	Makushtha	Mataki, Math	Moth, Mohar	Seeds
	42	*Vigna unguiculata* Walp. subsp. *cylindrica* (L.) Esel. [*Vigna cylindrica* (L.) Skeels]	Cow pea	Rah mash	Chavali	Rajmash, Chaura	Seeds
	43	*Vigna mungo* (L.) Hepper	Black gram	Mash	Udid	Urad, Udad	Seeds
	44	*Vigna radiata* (L.) Verdc.	Green gram	Mugda	Hirwe mug	Mag	Seeds
	45	*Vigna sinensis* (L.) Savi ex. Hassk.	Chinese beans		Chawli	Rajmash, Chaura	Seeds
	46	*Vigna trilobata* (L.) Verdc.	Wild moth bean	Mugdaparni	Ranmug	Ranmug	Seeds
	47	*Trigonella foenum graecum* L.	Fenugreek	Methika	Methi	Methi	All parts

No.	Family		Latin Name	English Name	Sanskrit Name	Marathi Name	Hindi Name	Parts Used
58	Fagaceae	1	*Quercus infectoria* Olivier	Gall oak, Dyer's oak	Majuphal, Mayaphal	Mayphal	Majuphal	Insect gall
59	Flacourtiaceae	1	*Flacourtia indica* (Burm.f.) Merr.	Govern's plum	Swadukantak	Kakare	Kantaie	Fruits
		2	*Hydnocarpus laurifolius* (Denn.) Sleumer	—	Turvak Katukapittha	Kadu kawith	—	Seed-oil
60	Fumariaceae	1	*Fumaria indica* (Haussk.) Pungsley	Fumitoru	Parpat	Pittapapda	Dhamgajaara	All parts
61	Gentianaceae	1	*Canscora decussata* (Roxb.) Schult. & Schult.	—	Shankhpushpi Dandotpala	Shankhahuli	Sankhahuli	All parts
		2	*Enicostema axillare* (Lam.) Reynal [*E. littorale* Bl.]	—	—	Mamejwa	Chotakirait	All parts
		3	*Swertia chirayita* Karsten	Chiretta	Kirat tikta	Kadechirait	Chiraita	All parts
		4	*Swertia densiflora* (Griseb.) Kashyapa [*Swerita decussata* Nimmo ex. Cl.]	—	—	Kadu	—	All parts
62	Hippocrateaceae	1	*Salacia chinensis* L.	—	Saptachakra	Ingali	Saptarangi	Roots
63	Hypoxidaceae	1	*Curculigo orchioides* Gaertn.	Black musale	Talmuli	Kali musali	Syah musali	Underground stem
64	Iridaceae	1	*Crocus sativus* L.	Saffron	Kunkumam	Keshar	Kesar, Jafran	Stamens
		2	*Iris germanica* L.	Orris root	Hemvati Shwet waacha	Balvekhand	Pohakar mul	Roots
65	Juglandaceae	1	*Juglans regia* L. var. *kumaonia* DC.	Walnut	Akshot	Aakrod	Akhrot	Bark, leaves, fruits
66	Lamiaceae (Labiatae)	1	*Anisomeles malabrica* R. Br.	Malabar catemint	Sprukka	Kapuri madhuri	Kaprimadhuri	Entire plant
		2	*Coleus amboinicus* Lour. [*C. aromaticus* Benth.]	—	—	Panancha owa	Patherchur	All parts
		3	*Coleus barbatus* (Andr.) Benth.	Coleus	—	Mainmul	—	Roots
		4	*Hyssopus officinalis* L.	Hyssop	—	—	Jufa	All parts
		5	*Leonotis nepetifolia* (L.) R.Br.	—	—	Matisul		Leaves, flowers
		6	*Leucus cephalotes* (Roth) Spreng.	—	Dronpushpi	Tumba	Guma	All parts
		7	*Majorana hortensis* Moench.	Sweet marjoram	Marubak	Marwq	Marwa	All parts

No.	Family		Latin Name	English Name	Sanskrit Name	Marathi Name	Hindi Name	Parts Used
		8	*Mentha arvensis* L.	Menthol	Putinash, Karpur	Pudina	Pudina	Oil crystals
		9	*Mentha spicata* L.	Menthol, Spear mint	Putinash	Pudina	Pudina	Oil crystals
		10	*Ocimum basilicum* L. var. *thysiflora* Benth.	Sweet basil	Berber	Subja, Rantulas	Subja	All parts
		11	*Ocimum gratissimum* L.	—	—	Ramtulas	Ramtulasi	Leaves
		12	*Ocimum tenuiflorum* L. [*O. sanctum* L.]	Sacred basil	Tulasi	Tulas Tulasi	Kala tulasi	All parts
		13	*Pogostemon parviflorus* Benth.	—	Phannijak	Phangala	—	All parts
67	Lauraceae	1	*Cinnamomum camphora* (L.) Presl.	Borneo camphor	Karpur	Kapur	Kapur, Baras Kapur	Leafwax
		2	*Cinnamomum tamala* Nees ex. Eberm.	Indian cinnamon	Tamalpatra	Tamalpatra	Tejpatta	Leaves
		3	*Cinnamomum verum* J.S. Presl. [*C. zeylani-cum* auct. non Bl.]	Cinnamon	Twak, Gudtwak	Dalchini	Dalchini	Bark, bark-oil
		4	*Litsea glutinosa* (Lour.) C.B. Robbins	Common fallow laurel	Medasak	Mendalkadi	Mendalkadi	Bark
68	Lecythidaceae	1	*Barringtonia acutangula* Gaertn.	—	Hijjal	Satphal, Samudraphal	Hijjal	Fruits
		2	*Barringtonia racemosa* Bl.	—	—	Samudrafal		Fruits
		3	*Careya arborea* Roxb.	—	Kumbhi	Kumbha	Kumbhi	Bark, flowers
69	Liliacceae	1	*Aloe vera* (L.) Burm f.	Aloe plant	Kumari, Ghruta kumari	Korfad	Garpatha	Leaf-juice
		2	*Asparagus racemosus* Willd. var. *javanicus* Baker	Wild Asparagus	Shatawari Narayani	Shatawari	Satawar	Roots
		3	*Asparagus adscendens* Roxb.	—	Shwet musali	Safed musali	Safed musali	Roots
		4	*Chlorophytum arundinaceum* Baker	—	Shwet musali	Safed musali	Safed musali	Roots
		5	*Colchium luteum* Baker	Sweet hermodactyl Colchicum	—	Suranjan	Suranjan	Corm
		6	*Drimia indica* (Roxb.) Jessop [*Urginea indica* (Roxb.) Kunth.]	Indian squill	Wanpal-andu	Rankanda, Kolkand	Jungali Pyaj	Bulb

No.	Family		Latin Name	English Name	Sanskrit Name	Marathi Name	Hindi Name	Parts Used
		7	*Gloriosa superba* L.	Glory lili, Superb lily	Langali, Kalihari Kalikarika	Kalalavi	Kalihari, Languli-kand	Rhizome
		8	*Scilla hyacinthina* (Roth) Mc Bride	—	—	Jungali kanda	Pahadi kand	Bulb
70	Linaceae	1	*Linum usitatissimum* L.	Linseed, common	Aatasi	Alasi, Javas	Alsi	Seeds
71	Lobeliaceae	1	*Lobelia nicotianaefolia* Heyne	Lobelia	—	Devnal, Boknal	Narsal	Seed-oil
72	Loranthaceae	1	*Dendrophthoe falcata* (L.f.) Ettings [*Loranthus falcatus* L.f.]	—	Baandak	Bandgul	Banda	Leaves, flowers
		2	*Viscum nepalense* Spreng, [*V.articulatum* auct.]	Mistle toe	—	Hadmod Hadmodi	Panpudu	Entire plant
73	Lythraceae	1	*Ammannia baccifera* L.	—	Agniwati, Agnipatri	Bharjam-bhul	Dadmari	All parts
		2	*Lagerstroemia parviflora* Roxb.	—	—	Lendi	Jarul	All parts
		3	*Lawsonia inermis* L. [*L. alba* Lamk.]	Henna, Heena	Mendika	Mehandi, Mendi	Heena	All parts
		4	*Woodfordia fruticosa* (L.) Kurz. [*W. floribunda* Salisb.]	—	Dhatki	Dhayati	Dhai	All parts
74	Magnoliaceae	1	*Michelia champaca* L.	—	Swarnacha-mpak	Sonchapha	Champa	Bark, flowers
75	Malvaceae	1	*Abelmoschus esculentus* (L.) Moench.	Lady's finger	Rishinda	Bhendi	Ramturai, Bhendi	Raw-fruits
		2	*Abelmoschus moschatus* Medik.	Musk mallow	Latakasturi	Kastur-bhendi	Mishka bhendi	Roots, leaves, seeds
		3	*Abutilon indicum* (L.) Sweet	Indian mallow	Atibala	Mudrika, Peteri, Karandi	Kangai	Bark, leaves, seeds
		4	*Althaea officinalis* L.	Marsh mallow	—	—	Khatmi	All parts
		5	*Gossypium arboreum* L.	Cotton	Uddyan karpas	Devkapus	Narm	Roots, leaves, seeds
		6	*Hibiscus cannabinus* L.	—	Ambashtha, Machika	Ambadi	Patson	Leaves, flowers
		7	*Hibiscus rosa-sinensis* L.	Shoe-flower	Japa	Jaswandi	Jasund	Flowers
		8	*Hibiscus syriacus* L.	Sharon	Japa	Nili jaswand	Jabakusum	Flowers
		9	*Kydia calycina* Roxb.	—	—	—	—	—
		10	*Sida acuta* Burm.f.	—	Nagbala	Tupkadi	Gangeran	Roots

No.	Family		Latin Name	English Name	Sanskrit Name	Marathi Name	Hindi Name	Parts Used
		11	*Side cordifolia* L.	—	Bala, Vattyalak	Bala, Chikana	Kharenti	All parts
		12	*Sida spinosa* L.	—	Kharyas-htika	Kante bale	Gulsakri	Roots, leaves
76	Meliaceae	1	*Aglaia elaeagnoidea* (Pers.) Benth	Perfumed cherry	Priyangu	Gahula	Uiyan	Seeds
		2	*Azadirachta indica* A. Juss.	Neem tree, Indian lilac	Nimb	Kadulimb, Nimb	Neem	All parts
		3	*Melia azedarach* L.	Persian lilac	Mahanimb	Limbara, Bakana nimb	Bakayan	Roots, bark, laeaves, flowers
		4	*Soymida febrifuga* (Roxb.) A. Juss.	—	Rohini	Rohan	Rohin	Bark
77	Menisper-maceae	1	*Cissampelos pareira* L. var. *hirsuta* (Ham. Ex. DC.) Forman	Velvet leaf	Laghu patha	Pahad wel	Path, Padhi	Entire plant
		2	*Cocculus hirsutus* (L.) Diels. [*C. villosus* DC.]	—	Vatsadani, Patal garudi	Wasanwel	Chirhata	Roots, leaves
		3	*Cyclea peltata* Burm.f.	Velvet leaf	Rai patha	Patha, Padawal	Rajpatha	All parts
		4	*Tinospora cordifolia* (Willd.) Miers ex. Hook.f. & Thoms	—	Guduchi	Gulvel	Giloy	All parts
78	Mimosaceae	1	*Acacia catechu* (Roxb. Ex Rottl.) Willd.	Black catechu	Khadir	Khair, Kath	Khair, Kattha	Heart-wood
		2	*Acacia chundra* (Roxb. Ex Rottl.) Willd.	Red catechu	Rakta Khadir	Lal khair	Lal kattha	Heart-wood
		3	*Acacia nilotica* (L.) Diels subsp. *indica* (Benth.) Brenan	Babool	Babhool, KInkirat	Babhool	Babool	Bark, gum
		4	*Acacia senegal* Willd.	Gum arabic	—	Dink	—	Gum
		5	*Albizia lebbeck* (L.) Willd	Parrot tree	Shirish, Shukapriya	Shirish	Siras	All parts
		6	*Albizia odoratissima* (L.f.) Benth.	Siris	Krishna shirish	Kala shirish	Kala siras	All parts
		7	*Dichrostachys cinerea* (L.) Wt. & Arn.	—	Veertaru	Sigamkati	Wellantar	Root, bark
		8	*Entada pursaetha* DC. [*E. scandens* auct. Non Benth]	—	Garbeej	Garambi	Gilgaccha	Wood, seeds
		9	*Mimosa pudica* L.	Sensitive plant	Lajjalu	Lajalu	Lajjawanti	Roots
		10	*Prosopis cineraria* (L.) Druce	—	Shami, Shktuphala	Shami	Shami	Bark, leaves, fruits

No.	Family		Latin Name	English Name	Sanskrit Name	Marathi Name	Hindi Name	Parts Used
79	Moraceae	1	*Artocarpus heterophyllus* Lam.	Jack fruit	Panas	Phanas	Kathal	Fruits
		2	*Ficus amplissima* J.E. Sm.	—	Plaksha	Pimpri, Pimparni	—	Bark
		3	*Ficus benghalensis* L.	Banyan tree	Byagrodh, War	Wad	Bur	All parts
		4	*Ficus carica* L.	Fig.	—	Anjeer	Anjeer	Fruits
		5	*Ficus hispida* L.f.	—	Kakodumbar	Bhui umber	Katgular	Bark, fruits
		6	*Ficus microcarpa* L.f. [*F. ratusa* auct. non L.]	—	Nandi wruksha	Nandruk	Weliya peepal	Roots, bark, fruits
		7	*Ficus racemosa* L. [*F. glomerata* Roxb.]	—	Udumbar	Umbar	Gular	All parts
		8	*Ficus religiosa* L.	Peepal	Ashwattha, Pippal	Pimpal	Peepal	Bark, leaves, fruits
		9	*Ficus rumphii* Bl.	—	Ahta	Ashta	Paakar	Bark, latex
		10	*Ficus virens* Ait.	—	Plaksha	Paeer	Pakhar	Bark, fruits
		11	*Morus alba* L.	Mulberry	Toot	Tutoo	Shahtoot	Fruits
80	Moringaceae	1	*Moringa oleifera* Lamk. [*M. pterigosperma* Gaertn.]	Horse radish, Drum stick	Shigru	Shwga	Sahijana	Roots, bark, leaves, seeds
81	Musaceae	1	*Musa paradisiaca* L.	Banana	Kadali	Kela	Kela	Rhizome, fruits
82	Myricaceae	1	*Myrica esculenta* Buch. Ham. ex. D. Don	Box myrtle, Bag berry	Katphal	Kaiphal	Kaiphal	Bark
83	Myristicaceae	1	*Myristica fragrans* Hout.f.	Nutmeg	Jatiphal	Jaiphal	Jaiphal	Seeds
84	Myrsinaceae	1	*Embelia ribes* Burm. f.	Babreng	Vidang, Krumighna	Wawding	Waywidang	Fruits
85	Myrtaceae	1	*Eucalyptus citriodora* Hook.	Lemon scented eucalypt	Tailaparni	Nilgiri	Neelgiri	Leaves, oil, gum
		2	*Eucalyptus globulus* Labill	Tasmanian blue gum	Tailaparni	Nilgiri	Neelgiri	Leaves, oil, gum
		3	*Psidium guayava* L.	Guava	Peruk, Peral	Peru	Amrud	Leaves, fruits
		4	*Syzygium aromaticum* (L.) Merr. & Perry	Clove	Lavang	Lavang	Laung	Flowers, buds
		5	*Syzygium cumini* (L.) Skeels	Black plum	Jambu	Jambhul	Jamun	Bark, leaves, fruits
86	Naucleaceae	1	*Haldina cordifolia* (Roxb.) Ridsd. [*Adina cordifolia* (Roxb.) Hook.]	Yellow teak	Haridru	Hed, Haldu	Haldu	Bark

No.	Family		Latin Name	English Name	Sanskrit Name	Marathi Name	Hindi Name	Parts Used
		2	*Neolamarckia chinensis* (Roxb.) Bosser (*Anthocephalus cadamba* Roxb.)	Wild cinchona	Kadamb	Kadamb	Kadam	Bark
87	Nelumbo-naceae	1	*Nelumbo nucifera* Gaertn. (*Nelumbium speciosum* Willd.)	Sacred lotus	Kamal, Padma, Pundric, Aravind	Kamal	Sahast radal, kanwal	All parts
88	Nyctaginaceae	1	*Boerhavia chinensis* (L.) Ascher. & Schweinf	Climbing hog weed	—	Mothi ghetuli	Bada Punarnava	All parts
		2	*Boerhavia diffusa* L.	Spreading hog weed	—	Ghetuli, Khapara	Lal Punarnava	All parts
		3	*Myrabilis jalapa* L.	Four 'O' clock plant	—	Sandhay-akali	Gulbas	Roots
89	Nymphaea-ceae	1	*Nymphaea alba* L.	White water lily	Shevt Kamal, Kumud	Pandhare Kamal	Safed Kamal, Kanodini	Flowers, seeds
		2	*Nymphaea nouchali* Burm.f. (*N. Stellata* Willd.)	Blue lotus	Nilkamal	Nile Kamal	Nelkamal	Flowers, seeds
		3	*Nymphaea pubesens* Willd.	Pink lotus	Raktakamal	Gulabi kamal	Lalkamal	Flowers. seeds
		4	*Nymphaea rubra* Roxb.	Red lotus	Rakta kamal	Lalkamal	Lalkasmal	Flowers, seeds
90	Oleaceae	1	*Jasminum auriculatum* Vahl.	—	Jtika	Juie	Juhi	Leaves, flowers
		2	*Jasminum calophyllum* Wall. Ex Clarke	—	—	—	—	Leaves
		3	*Jasminum grandiflorum* L. (*J. officinale* L. var. *grandiflorum* (L.) Bailey)	—	Jati	Jai, Chemali	Chambeli	Leaves, flowers
		4	*Jasminum officinale* L.	Poet's jasmin	Jati	Jaai, Chemali	Chambeli	Leaves, flowers
		5	*Jasminum multiflorum* (Burm.f.) Andr.	Kasturi Mallika	Kund	Kund	—	Leaves
		6	*Jasminum sambac* (L.) Ait.	Double Jasmine	Mallika	Mogra	Motia	Roots, leaves, flowers
		7	*Nyctanthes arbortristis* L.	—	Parijat	Prajaki	Harsindhar	Bark, leaves, seeds
		8	*Schrebera sweiteniodes* Roxb.	—	Mokshak, Ksharsh-reshta	Mokhada	Mokha, Eksira	—
91	Orchidaceae	1	*Eulophia nuda* Lindl.	—	Amarkand, Mankand, Malakand	Larunkand, Mankand	Amarkand, gorum	Pseudobulb
		2	*Orchis latifolia* L.	Salep	Munatak	Salam Mishri	Salam Mishri	Roots

No.	Family		Latin Name	English Name	Sanskrit Name	Marathi Name	Hindi Name	Parts Used
		3	*Vanda tessellata* (Roxb) Hook. ex G. Don	—	Vrikshadani	Bande	Banda	Roots
92	Oxalidaceae	1	*Biophytum sensitium* (L.) DC.	—	—	Lajari	Lakjan	All parts
		2	*Oxalis corniculata* L.	Indian sorrel	Changeri	Ambati, Bhusaroati	Tipati chuka	All parts
93	Pandanaceae	1	*Pandanus odoratissimus* L. (=*P. fascicularis* Lamk.)	Screw pine	Suvarna Ketaki	Kevada	Keora	Leaves, Flowers
94	Papaveraceae	1	*Argemone mexicana* L.	Mexican poppy	Satyanashi	Pivia Dhotra	Bhadbhand	All parts, latex
		2	*Papaver somniferum* Dunn.	Opium poppy	Ahifen	Aphoo	Aphoo	Latex, fruits
95	Parmeliaceae	1	*Parmelia perlata* (Huds.) Ach	Stone flower	Shaileya	Dagadpuhl	Chhadila	Entire plant
96	Pedaliaceae	1	*Pedalium murex* L.	—	Bruhad, Gokshuru	Mother Gokhru	Faridbuti	Roots, fruits
		2	*Sesamum orientale* L. (=*S. indicum* L.)	Sesame, Gingelly seed	Tilam	Til	Til	Seed-oil
97	Periplocaceae	1	*Cryptolepis buchanani* Roem. & Schult.	—	Shweta Sariwa	Pandhri Kawali	Kalisar	Roots
		2	*Hemidesmus indicus* (L.) Schult.	Indian sarsaparilla	Sariwa	Upalsari	Anantmul	Roots
98	Pinaceae	1	*Cedrus deodara* Roxb.	Himalayan Cedar	Deodar	Deodar	Deodar	Wood
		2	*Pinus roxburghii* Sar. (=*P. longifolia* Roxb.)	Long leaved pine, Chir pine	Saral	Suchiparn	Cheer	Wood, resin
99	Piperaceae	1	*Pipper betle* L.	Bettle leaf	Nagvalli	Nagvel	Pan	Leaves, fruits
		2	*Piper cubeba* L.	Cubebs	Kankol	Chanakabab	Kababohini	Raw fruits
		3	*Piper longum* L.	Long pepper	Pippali	Lendi Pimpli	Peeper	Roots, fruits
		4	*Piper nigrum* L.	Common pepper, (Black)	Marich	Mire	Kali Mirch	Fruits
100	Pistaciaceae	1	*Pistacia vera* L.	Pistachio	Mukulak	Pista	Pista	Insect galls on leaves, seeds
101	Plantaginaceae	1	*Plantago ovata* Forsk.	Spongel seed	Ishat Goal, Ashwakarn Beej	Isabgoad	Isabgul	Seeds
102	Plumbaginaceae	1	*Plumbago auriculata* Lam. (*P. capensis* Thunb.)	Blue lead wort	Neel Chitrak	Nila Chitrak	Nil Chitrak	Roots

No.	Family		Latin Name	English Name	Sanskrit Name	Marathi Name	Hindi Name	Parts Used
		2	*Plumbago indica* L. (*P. rosea* L.)	Red lead wort	Rakta Chitrak	Lal Chitrak	Lal Chitrak	Roots
		3	*Plumbago zeylanica* L.	Black lead wort	Chitrak, Anal, Vayal	Chitrak	Chitrak	Roots
103	Poaceae (Gramineae)	1	*Arundo donax* L.	—	—	Devnal	—	Stem, leaves
		2	*Cymbopogon citratus* (DC.) Stapf	Lemon grass	Bhutik, Sugandh Bhutrun	Gavati chaha	Gandha-bena, Hari Chay	Leaves-oil
		3	*Cymbopogon martini* (Roxb.) Wats.	Rosha Grass	Rohin	Tikhadi, Rosha Gavat	Motiya Ghas	Oil
		4	*Cymbopogon winterianus* Jowett	Citronella	Kutrun	Sitronela, Takkatan	Ganjani	Oil
		5	*Cynodon dactylon* (L.) Pers.	Dog's teeth grass	Durva, Shatvirya, Sahas-travirya	Durva, Harali	Hariyali, Haridub	Entire plant
		6	*Desmostachya bipinnata* (L.) Stapf	—	Kush, Darbha	Darbha	Kush, Dabha	Roots
		7	*Saccharum arundina-ceum* Retz. (*S. munja* roxb., *S. moonja* Royel)	—	Shar, Ghadr-amujja	Shar	Ramsar, Bhadramunj	Leaves
		8	*Saccharum officinarum* L.	Sugarcane	Ekshu	Usa	Ganna	Stem juice, sugar
		9	*Saccharum spontaneum* L.	Wild sugar cane	Kasekshu	Kasai	Kas	Leaves
		10	*Vetiveria zizanioides* (L.) Nash	Khus, Cuscus grass	Peetmula, Amlaparni	Rewanchini, Rewachini	Rewand-chini	Roots
104	Polygonaceae	1	*Rheum emodi* Wall.	Rhubarb	Peetmula, Amlaparni	Rewanchini, Rewachini	Rewand-chini	Roots
		2	*Rumex vesicarius* L.	Country sorrel	Chukrika	Ambar	Chuka	Leaves
105	Portulacaceae	1	*Portulaca oleracea* L.	Garden Purslane	Bruhlloni	Ghol	Badi Lona	Stem, leaves
		2	*Portulaca quadrifida* L.	—	Lona, Loni	Lahan Gohl	Choti Lona	Stem, leaves
106	Ranuncu-laceae	1	*Aconitum heterophyllym* Wall. (*A. ferox* Wall.)	Indian atees	Atiwish	Atiwish	Atisa	Roots
		2	*Aconitum nepellus* L.	Indian Aconite	Wastanabh	Bachnag	Wachanag	New roots
		3	*Clematis heynei* M.A. Rao (*C. triloba* Heyne ex Roth)	—	Murva	Morvel, Ranjai	Jastor	All parts
		4	*Delphinium denudatum* Wall.	—	Nirvisha	Nirvish	Nirvishi Jadwar	Roots

No.	Family		Latin Name	English Name	Sanskrit Name	Marathi Name	Hindi Name	Parts Used
		5	*Nigella sativa* L.	Black cumin	Upakun-chika	Kale jire	Kalounji, Jaldhaniya	Seeds
		6	*Ranunculus sceleratus* L.	—	Kandir	—	Jaldhaniya	—
107	Rhamnaceae	1	*Ziziphus jujuba* Mill.	Common Jujube	Sauvir, Rajbadar	Bor	Unnab	Bark, fruits
		2	*Ziziphus mauritiana* Lamk. (*Z. jujuba* auct. non Lam.)	Cottony Jujube	Kol, Badar	Bor	Ber	Bark, fruits
108	Rosaceae	1	*Prunus dulcis* (Mill) D.A. Webb. (*P. amygdalus* Batsch)	Almond	Vatad, Netropa-maphala	Badam	Badam	Seeds
		2	*Prunus cerasoides* D. Don (*P. puddum* Wll) Roxb. Ex Brandis	Himalayan Cherry	Padmak	Padmak-ashath	Padmak	Branches
		3	*Rosa centifolia* L.	Rose	Shatpatri Taruni	Gulab	Gulab	Flowers-buds
109	Rubiaceae	1	*Catunaregam spinosa* (Thunb.) Tirven. (*Randia dumetorum* (Roxb.)	Emetic nuttree	Madan	Gelphal	Meinphal	Fruits
		2	*Gardenia gummifera* L.f.	—	Nadihidgu	Dikemali	Dikamali	Resin
		3	*Ixora coccinea* L.	—	—	Pankul, Patkali	—	Roots
		4	*Paederia foefida* L.	—	Prasarini	Prasarani	Prasarani	All parts
		5	*Rubia cordifolia* L.	Maddar root	Manjishtha	Manjishtha	Manjitha	Roots
		6	*Spermadictyon suaveolens* Roxb.	—	—	Jitsaya	Jitsaya	Roots
110	Rutaceae	1	*Aegle marmelos* (L.) Corr.	—	Bilwa	Bel	Bel	All parts
		2	*Citrus aurantifolia* (Christm.) Swingle	Lemon, Lime	Nimbuk	Kagadi Limbu	Ninbu	Fruits
		3	*Citrus limon* L.	Lemon	Dantshath	Eedlimbu	Jmbiri Nimbu	Fruits
		4	*Citrus medica* L.	—	Matulung	Mahalumg	Bijaura	All parts
		5	*Ruta chalepensis* L. (*Ruta graveolens* L.var *angustifolia* Hook f.)	Garden Rue	—	Satap	Sudab	All parts, Oil
		6	*Zanthoxylum armatum* DC.	Toothache tree	Tumburu Tejswini	Nepali Dhaniya	Tejbal	Fruits
111	Salvadoraceae	1	*Salvadora persica* L. var. *wightiana* (Planch. ex thw.) Verdc	—	Pilu	Pilu, Khakan	Kharjal	All parts

No.	Family		Latin Name	English Name	Sanskrit Name	Marathi Name	Hindi Name	Parts Used
112	Samydaceae	1	*Casearia elliptica* Willd. (*C. tomentosa* Roxb.)	—	—	Massi	Chilla, Bairi	All parts
		2	*Casearia esculenta* Roxb.	—	—	—	Chilla, Bairi	Roots, leaves
113	Santalaceae	1	*Santalum album* L.	Sandal wood	Shrikhand Chandan	Chandan	Safed Chandan	Heart-wood, oil
114	Sapindaceae	1	*Cardiospermum helicacabum* L.	Heart pea	Karnsfota	Kanfuti, Kapalfodi	Kanfat	All parts
		2	*Sapindus emarginatus* Vahl.	Soap nut	Arishtak	Ritha	Ritha	Fruits
		3	*Sapindus laurifolius* Vahl.	Soap nut	Arishtak	Ritha	Ritha	Fruits
		4	*Schleichera oleosa* (Lour.) Oken (*S. trijuga* Willd.)	Ceylon oak	Koshamra	Kusumb, Kosimb	Kosam	Bark, seed-oil
115	Sapotaceae	1	*Madhuca longifolia* (Koen.) Macbr.	Mowar, Butter Tree	—	—	S. Indian Mahua	Seed-oil
		1a	*Madhuca indica* Gmel.	Illipe	Madhuk	Moh, Mohada	Mahua	Flowers
		2	*Manilkara hexandra* (Roxb.) Dub. (*Mimusops hexandra* Roxb.)	—	—	—	—	—
		3	*Manilkara zapota* (L.) P. van Royen (*Achras sapota* L.)	Sapota	—	Chiku	Sapeta, Chiku	Bark, seeds
		4	*Mimusops elengi* L.	—	Bakul	Bakul	Maulsari	Bark, flowers, fruits
116	Saxifragaceae	1	*Bergenia ligulata* (Wall.) Engl.	—	Pashan Bhed	Pashan Bhed	Pakhat bhed, Silada	Exudate
117	Scrophu-lariaceae	1	*Bacopa monnieri* (L.) Wettst.	Thyme leaved gratiola	Brahmi, Toyvalli	Brahmi	Barami, Brahmi	Entire plant
		2	*Digitalis purpurea* L.	Digitalis, foxglove	Tilpushpi Dhumra-pushpi	—	—	Leaves
		3	*Limnophila indica* (L.) Druce	—	Ambuj, Amrag-andha	Ambulli	Kuma	Entire plant
		4	*Picrorhyza scrophular-iiflora* Pennell (*P. kurroa* auct. Non Royle)	Picrorhiza	Katurohini	Kali Kutki	Kutaki, Katuki	Roots
118	Simarou-baceae	1	*Ailanthus exselsa* Roxb.	—	Mahanimb, aralu	Maharukh	Ghodanib	Bark
119	Smilacaceae	1	*Smilax lanceifolia* Roxb.	China root	Dwipantar Wacha	Chopachini	Chopachini	Roots
		2	*Smilax zeylanica* L.	Wild sarsa-parilla	Badarvalli Ushba	Dhorvel, Jangali Ushba	Ramdatun Ushba magrabi	Roots

No.	Family		Latin Name	English Name	Sanskrit Name	Marathi Name	Hindi Name	Parts Used
120	Solanaceae	1	*Atropa belladona* L.	Belladona	Krushna Phala	Suchi	Sag Angoor	Roots, leaves
		2	*Datura metel* L.	Thorn apple	Krushna Dhattur	Kala Dhotra	Kala Dhattura	All parts
		3	*Hyoscyamus niger* L.	Henbane	Paarsik Yavani	Khurasni Onwa	Khorasani Ajavayan	Leaves, flowers, seeds
		4	*Nicotiana tabacum* L.	Tobacco	Tamrakut	Tambaku	Tamakhu	Leaves
		5	*Solanum anguivi* Lamk (*S. indicum* auct. non L.)	Indian night shade	Bruhati	Dorali, Dorale	Badi Kateri	Roots, fruits, seeds
		6	*Solanum nigrum* L.	Black night	Kakmaachi	Kamoni	Makoy	All parts
		7	*Solanum virginianum* L. (*S. xanthocarpum* Schrad. Et Wendle)	Yellow berried night shade	Kantkari	Ringani, Bhui Ringani	Bhatak-taiya, Kateri	All parts
		8	*Withania somnifera* (L.) Dunn. var. *ashwagandha* Cl. var. *somnifera*	—	Ashwa-gandha	Askand Dnorgunja	Asgandh Askand	Roots, seeds Roots
121	Sterculiaceae	1	*Abroma augusta* (L.) L.f.	Devil's cotton, Perennial Indian hemp	Pishack Karpas	—	Ulat Kambal	Roots, leaves
		2	*Helicteres isora* L.	—	Awartani	Murads-henga	Maradasingi, Marorphali	Roots, fruits
		3	*Pterospermum acerifolium* Willd.	—	Muchakund Karnikkar	Kaniyar	Kaniar Karchepi	Bark, leaves, flowers
		4	*Pterospermum suberifolium* Lamk.	—	Muchakund	Muchakund	Mucha-kund	Flowers
		5	*Sterculia foetida* L.	—	—	Deshi Badam, Punai	Poon	Bark
		6	*Sterculia urens* Roxb.	Karaya Gum	—	Kandol, Sardhol	Gulu	Gum
122	Strychnaceae	1	*Strychnos nux- vomica* L.	Poisonnut	Vishtinduk, Kaktinduk	Kajara	Kuchala	Bark, seeds
		2	*Strychnos potatorum* L.f.	Clearing nut	Path Pras-adini Katak	Nirmali	Nirmali	Seeds
123	Symplocaceae	1	*Symplocos racemosa* Roxb.	Lodh	Lodhra	Lodhra	Lodha	Bark
124	Taxaceae	1	*Taxus baccata* L.	Himalayan yew	Talispatra	Talispatra	Talispatra Barmi	Leaves
125	Thymel-aeaceae	1	*Aquillaria agallocha* Roxb.	Eagle wood	Aguru, Krum-ijagdha	Agar	Agar	Wood

No.	Family		Latin Name	English Name	Sanskrit Name	Marathi Name	Hindi Name	Parts Used
126	Tiliaceae	1	*Corchorus capsularis* L.	Jute	Karshak, Kahudra-chanchu	Chonche	Naricha	Leaves, seeds
		2	*Corchorus fascicularis* Lamk.	—	—	Harankhuri, Bafali	Mafali	All parts
		3	*Corchorus olitouius* L.	—	Bruhch-anchu	Mothe Chonche	Banpat	Leaves, seeds
		4	*Corchorus trilocularis* L.	—	Katucha-nchu	Kadu Chonchu	Pat	Seeds
		5	*Elaeocarpus sphaericus* (Gaetn.) K. Schum. (*Elaeocarpus ganitrus* Roxb.)	—	Rudraksha	Rudraksha	Rudraksha	Fruits
		6	*Grewia asiatica* L.	—	Parushak	Phalasa	Phalasa	Fruits
		7	*Grewia polygama* Roxb.	—	—	Govali	Kukurbicha	Leaves
		8	*Grewia scabrophylla* Roxb.	—	—	Pandhari Dhaman	Saphed Dhamani	Flowers
		9	*Grewia tiliifolia* Vahl.	—	Dhanvan, Dharman	Dhaman	Dhamani	Bark
		10	*Grewia villosa* Willd.	—		Kharmati	Panja	Bark
127	Trapaceae	1	*Trapa natans* Roxb. subsp. *Bispinosa* (Roxb.) Makino	Water chestnut	Shrugatak	Shingada, Shengada	Singada, Sindhara	Fruits
128	Ulmaceae	1	*Holoptelea integrifolia* Roxb.	—	Chirbilva	Vavali, Vaval	Chirmil	Roots, bark, leaves, seeds
129	Valerianaceae	1	*Nardostachys grandiflora* DC. (*Nardostachys jatamansi* DC.)	Spikenard	Jatamansi	Jaatamansi	Jatamansi, Balchad	Under-ground stem
130	Verbenaceae	1	*Avicennia officinalis* L.	White mangrove	Tuvrak	Tivar	Bina	Bark, seeds
		2	*Callicarpa tomentosa* (L.) Murray (*C. lanata* L.)	—	—	Aisar, Yesar	Bastra	Bark, leaves
		3	*Clerodendrum inerme* (L.) Gaerth.	—	Kundali, Samudra Yuthika	Vanjai, Koyanel	Sangkupi	Roots, leaves
		4	*Clerodendrum multiflorum* (Brum.f.) O. ktze.	—	—	Hajar Mogara	—	—
		5	*Clerodendrum serratum* (L.) Moon	—	Bharli, Bhu-mijambu	Bhaarang?	Bharangi	Roots
		6	*Clerodendrum viscosum* Vent.	—	Kari, Kari, Bhandir	Bhandir	Bhant	Roots, leaves
		7	*Gmelina arborea* Roxb.	—	Gambhri	Shivan	Gambhar	Roots, leaves fruits

No.	Family		Latin Name	English Name	Sanskrit Name	Marathi Name	Hindi Name	Parts Used
		8	*Lantana camara* L. var. *aculeata* (L.) Mold.	—	—	Ghaneri	—	Fruits
		9	*Phyla nodiflora* (L.) Green	—	Vashir	Jalpimpali	Bhuiokara	All parts
		10	*Premna herbacea* Roxb.	—	—	Bharang	Barangi	Roots
		11	*Premna serratifolia* L. (*Premna integrifolia* L.)	—	Arani, Agnimanth	Nakhel, Arani	Arni, Ageth	Roots, leaves
		12	*Tectona grandis* L.f.	Teak	Sak	Sagvan	Sagun	Leaves, flowers, seeds
		13	*Vitex agnus- castus* L.	Chaste tree	Renuka	Renuka beej	Samhalu	Fruits
		14	*Vitex negundo* L.	Three leaved chaste tree	Nirgundi, Sinduvar	Nirgundi	Samhalu	All parts
131	Violaceae	1	*Hybanthus enneaspermus* (L.) F. Muell.	—	Purushratna	Ratan Purush	Ratan Baras	All parts
		2	*Viola odorata* L.	Sweet violet	Kampushpa	Kampushpa	Banafshah	All parts
132	Vitaceae	1	*Cayratia trifolia* (L.) Domin	Fox gape	Amlaparni	Ambat wel	Amal vel	Roots, leaves
		2	*Cissus adnata* L.	—	—	Naden	Kolezen	Root-tubers
		3	*Cissus quadrangula* L. (*C. quadranguldris* auct. Mult.)	—	Asthisa-ndhan	Kandwel	Harjor	Old stem
		4	*Cissus setosa* Roxb.	—	—	Khaj goli	Harval	All parts
		5	*Leea crispa* Van Royen ex L.	—	—	—	Namchelt	Roots, leaves
		6	*Leea indica* (Burm.f.) Merr.	—	Kurkur Jivha	Karkani	—	Roots, bark
		7	*Leea macrophylla* Roxb.ex Hornem	—	Dhol Samudrika	Dinda	Samudrak	Roots
		8	*Vitis vinifera* L.	Grape wine	Draksha	Drakshe	Angur	All parts
133	Zingiberaceae	1	*Alpinia galanga* (L.) Willd.	The greater galangal	Kuljjan	Koshtha Kulanjan	Bada Kulinjan	Rhizome
		2	*Alpinia officinarum* Roxb.	The lesser galangal	Chinwacha	Kulinjan	Kulinjan	Roots
		3	*Amomum aromaticum* Roxb.	—	—	—	—	—
		4	*Curcuma amada* Roxb.	Mango-ginger	Amra haridra	Ambehalad	Amahaldi	Rhizome, root-tubers
		5	*Curcuma angustifolia* Roxb.	Indian arrow root	Tawkshir	Tawkhir	Tikhur	Rhizome
		6	*Curcuma aromatica* Salisb.	Wild turmeric	Banharidra	Ran halad, Ambe halad	Banhaldi	Rhizome, root-tubers
		7	*Curcuma caesia* Roxb.	Black curcuma	Krishna Haridra	Kali Halad	Kali Haldi	Rhizome

No.	Family		Latin Name	English Name	Sanskrit Name	Marathi Name	Hindi Name	Parts Used
		8	*Curcuma longa* L.	Turmeric	Haridra	Halad	Haldi	Rhizome, root-tuber
		9	*Costus speciosus* (Koen.) Sm.	—	Kemuk, Peu	Pew	Kebua	Rhizome
		10	*Elettaria cardamomum* Maton	Malabar cardamum	Ela	Weldoda	Choti Elachi	Fruits, seeds
		11	*Kaempferia galanga* L.	—	Chandr-mulik	Kachari, Chandmul	Chandmul	Rhizome
		12	*Kaempferia rotunda* L.	—	Bhumic-hampak	Bhuichapha	Bhumi-champa	Rhizome
		13	*Zingiber officinale* Rosc.	Ginger	Ardrak	Ale	Adrak, Adarak	Rhizome
		14	*Zingiber purpureum* Roxb.	Wild ginger	Banrdrak	Pivale koshtha	Banada	Rhizome
		15	*Zingiber zerumbet* Sm.	—	Kulanj	Kulanj	—	Rhizome
134	Zygophy-llaceae	1	*Fagonia cretica* L.	Khorasan-thorn	Dhanvayas	Dhamasa	Hingua-dramahar	All parts
		2	*Guaiacum officinale* L.	Gum guaiacum	Jivdaru	Lohalakkad	Giaku	Wood, gum
		3	*Tribulus alatus* Del.	Winged caltrap	—	—	Gojhure-kalan	Fruits
		4	*Tribulus terrestris* L.	Small caltrap	Kshudra-gokshur	Sarate	Chota Gokharu	Roots, fruits